TREASURY OF *Audubon Birds*

TREASURY OF *Audubon Birds*

130 PLATES *from* THE BIRDS OF AMERICA

JOHN JAMES AUDUBON

Introduction by

ALAN WEISSMAN

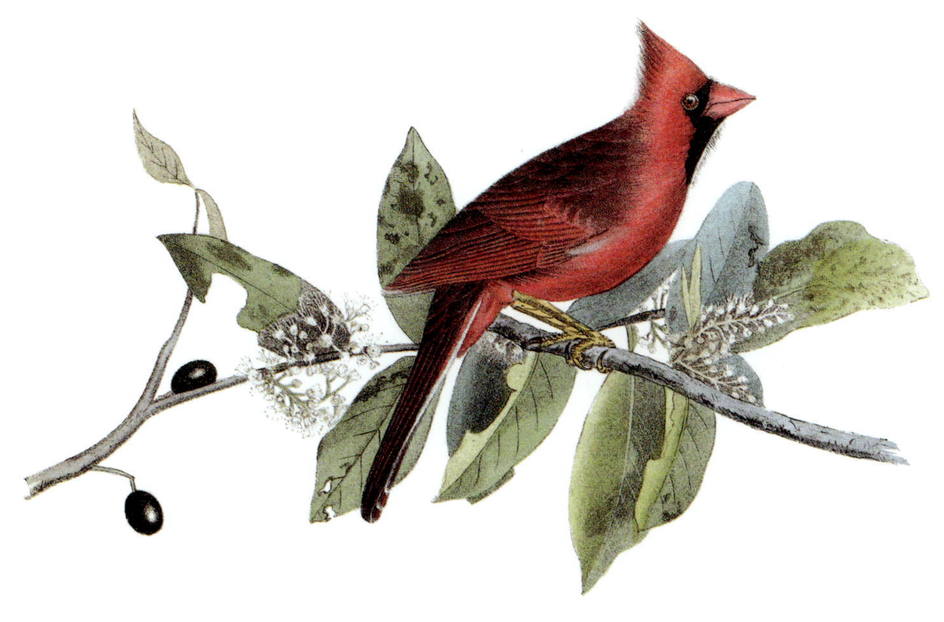

DOVER PUBLICATIONS

Garden City, New York

Bibliographical Note

Treasury of Audubon Birds: 130 Plates from The Birds of America is a new selection, first published by Dover Publications in 2020, of 130 plates from *The Birds of America, from Drawings Made in the United States and Their Territories,* by John James Audubon, F. R. SS. L. & E., first published in New York by J. J. Audubon and in Philadelphia by J. B. Chevalier from 1840 through 1844. The Introduction by Alan Weissman originally appeared in *Treasury of Audubon Birds in Full Color: 224 Plates from The Birds of America,* published by Dover Publications New York, in 1993.

Library of Congress Cataloging-in-Publication Data

Names: Audubon, John James, 1785–1851, author.
Title: Treasury of Audubon birds : 130 plates from the Birds of America / John James Audubon ; introduction by Alan Weissman.
Other titles: Birds of America. Selections
Description: Garden City, New York : Dover Publications 2020. | Summary: "Featuring the snowy egret, wild turkey, brown pelican, screech owl, and many others, this new collection gathers 130 select plates from Audubon's Octavo edition. Includes an informative Introduction to the artist and his work"—Provided by publisher.
Identifiers: LCCN 2019040556 | ISBN 9780486841793 (paperback) | ISBN 0486841790
Subjects: LCSH: Birds—North America—Pictorial works. | Birds in art.
Classification: LCC QL681 .A972 2020 | DDC 598.097—dc23
LC record available at https://lccn.loc.gov/2019040556

Manufactured in the United States of America
84179004 2022
www.doverpublications.com

List of Plates

The current common name of the bird is the first name given. The name by which the bird was known to John James Audubon, if it's different, follows in square brackets. The current scientific name is in parentheses.

Introduction

John James Audubon today is most often associated with the conservation-oriented societies that bear his name—not that his achievements as a naturalist, a man of letters, and even a pioneer are forgotten. Primarily, though, Audubon was an artist, and his greatest legacy is his monumental series of portraits of the birds of the United States of America. For more than a half-century, popular reproductions of these vital prints have created widespread awareness of Audubon's most enduring accomplishment.

Much less recognized—if not entirely obscure—to most viewers of this superb art is the somewhat complicated relationship between Audubon's original artistic endeavor and the prints that most people see.

Born in what is now Haiti in 1785, Audubon was educated in France, his paternal homeland. He studied both natural science and art, a background that made him ideally equipped to observe and interpret nature. He never failed to do this at least informally, for nature had been a passion since his early childhood. When, still young and enthusiastic, he arrived on his father's Pennsylvania estate in 1803, he began a love affair with North American wildlife, particularly birds, that ended only with his death in 1851. Through the many years when he pursued various unsuccessful business ventures, many of them in what was then only a semitamed wilderness in the interior of the continent, he always found time to observe the winged creatures of the woods, fields, and waterways—and to shoot them, to the dismay of many conservationists today. Audubon lived before any distinction was made between studying wild creatures and killing them. His conservationist's point of view—mild by modern standards, though remarkable for his time—developed gradually, as he lamented the wanton slaughter of avian life that he observed in his widespread travels.

In 1820, Audubon finally focused on his life's work, the ambitious project of painting *all* the birds then known in America. The difficulties of travel in those days posed a formidable obstacle, and many of the paintings had to be based on preserved specimens sent to him by others. Still, Audubon spent many years trekking through wilderness, observing most of his subjects firsthand on their breeding or wintering grounds or in migration. Without the data we have today, it was tricky to pinpoint the location of a migrating species at a given time of year. As he pursued this arduous task, obtaining many specimens, he recorded in voluminous journals marvelous accounts of bird behavior never before seen—except, perhaps, by Native Americans. In many instances, the encroachments of civilization on bird populations and habitats would make duplicating this achievement impossible in only a few years. Audubon also augmented his records with colorful, exciting accounts of his often perilous adventures. The resulting output of words and pictures was a unique, enduring tribute to his adopted country as well as a major contribution to ornithological art, far surpassing in quality anything of the kind that had been done before. Audubon's only real rival, Alexander Wilson (1766–1813), who for years worked along the same lines, was the superior ornithologist but distinctly inferior to Audubon as an artist.

Audubon mostly used watercolors but also pastels, ink, and other media. He engaged the assistance of others for the plants and the backgrounds; a few birds were painted by his younger son, John Woodhouse Audubon. The paintings took long to produce, but this was only the first part of the plan. Audubon arranged to have engravings made from the paintings, which would enable many copies to be produced and colored in. This was very expensive, as each copy of each painting had to be individually hand-colored. To finance the undertaking, Audubon expended a great deal of energy promoting subscriptions to his masterwork among those wealthy enough to afford it.

Once he had sufficient financial backing, Audubon took special care to select the right engraver to reproduce his art. After considerable difficulty and a major false start, the job fell to English engraver Robert Havell, Jr., one of the world's finest artists of the kind. One of the unusual characteristics of the first edition of *The Birds of America*—known as the "double-elephant folio"—was that its 435 plates reproduced their subjects life-size. Even with the enormous dimensions of 29½ by 39½ inches, this necessitated presenting some large birds in twisted, unnatural positions, a quirk of Audubon's art that subsequent artists were wise not to imitate. Despite this and a few other idiosyncrasies, the results were magnificent, more so in view of the labor involved. Havell first made copperplate engravings of the paintings; then as many as fifty assistants painstakingly, by hand, colored each print of each engraving. The tremendous amount of work involved explains the size of Havell's staff, as well as the price of each final set: $1,000, a gigantic sum at that time. In fact, Audubon had considerable difficulty collecting the money from even some of his wealthiest subscribers! With his own time-consuming field research, the actual painting, the delivery of the paintings to England, and the process of engraving and coloring, it is little wonder that the publication of the full double-elephant folio encompassed more than eleven years, from 1827 to 1838. Considering that concurrently Audubon took voluminous notes and submitted them for separate publication as *Ornithological Biography*, its 3,000 pages in five volumes a massive accomplishment in itself, the wonder is that he did it all in *only* eleven years. As with the paintings, he did receive assistance—he was only human—this time from Scottish-born scientist William MacGillivray.

With all this, Audubon was still unsatisfied. He wanted his art to appear in a more affordable edition that could be enjoyed by a wider audience. Thus, he conceived of the octavo edition, which soon began to appear, in 1840 through 1844, in 100 parts, each with five plates and text adapted from *Ornithological Biography.* The whole was finally sold as a complete set of seven volumes. Apart from size—the octavo plates measured about 6½ by 9¾ inches, obviously much less cumbersome than those of the double-elephant folio—there are other differences. The octavo added eleven new species—or birds Audubon believed to represent new species. There were more than eleven new plates, however, since different species occupying the same plate in the double-elephant folio were separated, like the Chestnut-Backed and Black-Capped Chickadees on plates 41 and 42 in this book. Backgrounds and other, usually subsidiary, details were sometimes different as well.

Another difference was the method of reproduction. For the octavo, the revised and augmented original plates were copied in reduced form, by means of a camera lucida, by John Woodhouse Audubon. The color reproduction was by lithography, assuring greater consistency, as well as speed, than with the laboriously hand-colored plates of the double-elephant folio.

It is from this octavo edition that the present selection has been made. The arrangement is very roughly that of reverse taxonomic order, that is, from finches to loons, with a few exceptions made for reasons of position. The most radical and potentially misleading displacement is of the Limpkin, which is more closely related to the cranes than to the herons it now appears with. On the plates, the birds are identified by their current common names. In the List of Plates, one of Audubon's names, where significantly different from the common name used today, follows, and then the current scientific name. Audubon often applied several names to the same bird in different contexts.

There is no doubt that Audubon will always be an inspiration to nature lovers and conservationists. Yet his masterwork, *The Birds of America*, is the achievement by which he most wished to be remembered. Nearly two centuries later, we may well ask how valuable this work remains, apart from its simple historical precedence.

Some of it surely will last because Audubon was in the right place at the right time. Much of his exploration of the continent came just before relentless land-clearing and development and market hunting forever altered the numbers and distribution of North American bird populations. Simply for his having documented the habits and appearance of now-extinct species like the Passenger Pigeon and the Carolina Parakeet, we must be forever grateful.

There is an even more valuable dimension of his achievement. It is true that nature artists with greater technical facility have since appeared and much of Audubon's ornithological data have been superseded. Sometimes, where he was pressed for time or had not seen a live bird in its natural habitat, his rendering was not very accurate or inspired. But when he thoroughly immersed himself in his subject, the result has never been surpassed and rarely equaled.

Examine, for example, the plate of the Great Carolina Wren (plate 37). This was based on a painting done in Louisiana or Mississippi in 1822; the plant, a scarlet buckeye, was painted by Audubon's assistant, Joseph Mason. All who have observed this bird will appreciate the characteristic attitude of the male singing energetically from the top of the plant. Audubon's own words complement his picture:

> The quickness of the motions of this active little bird is fully equal to that of the mouse. Like the latter, it appears and is out of sight in a moment, peeps into a crevice, passes rapidly through it, and shews itself at a different place the next instant. When satiated with food, or fatigued with these multiple exertions, the little fellow stops, droops its tail, and sings with great energy a short ditty something resembling the words *come-to-me, come-to-me*, repeated several times in quick succession, so loud, and yet so mellow, that it is always agreeable to listen to them.

Here, where his brush and pen preserved what his eye and ear had seized in the intensity of his direct experience, Audubon is at his best. This, surely, is the Audubon that will longest endure.

ALAN WEISSMAN

TREASURY OF
Audubon Birds

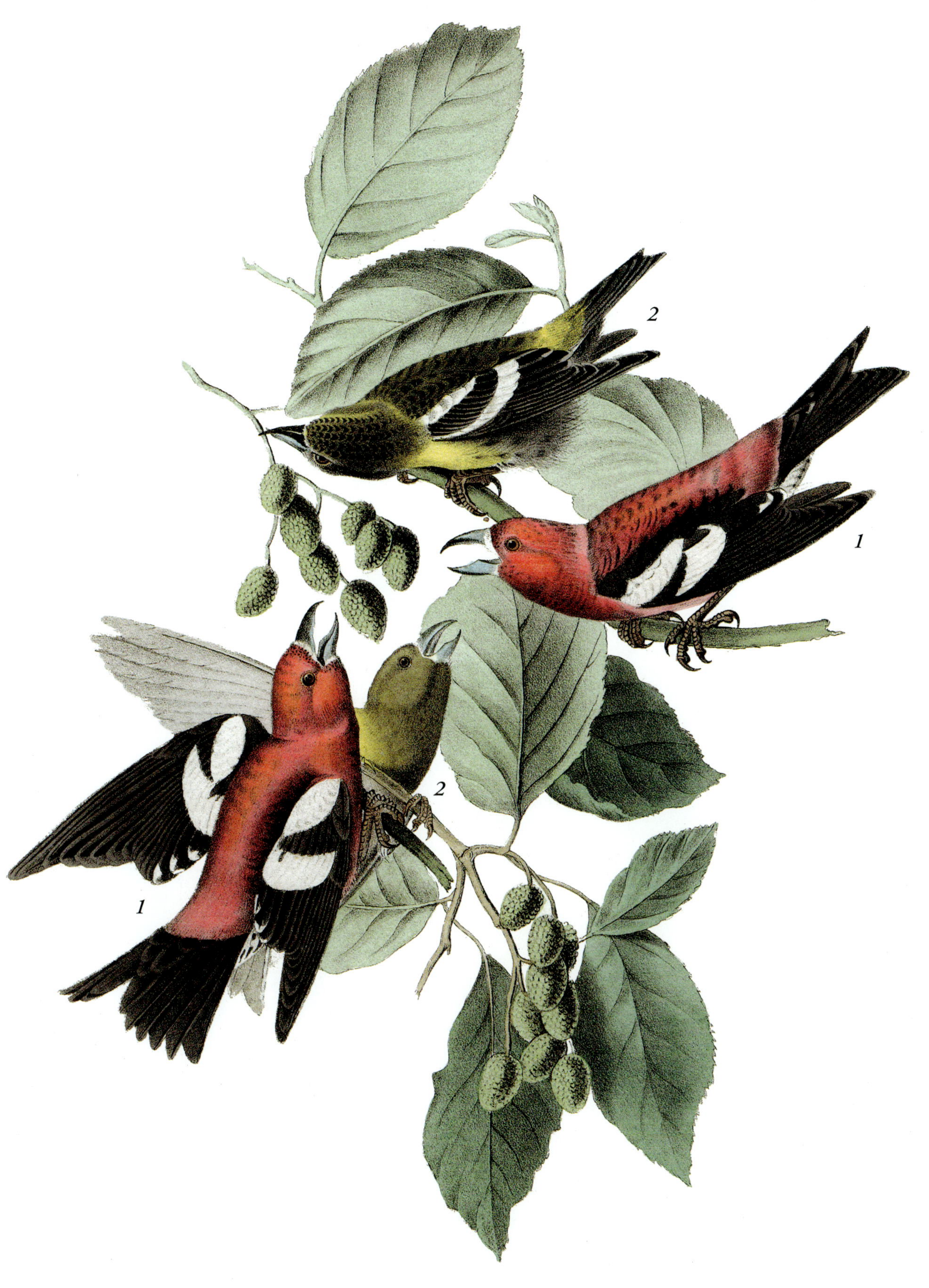

1. White-Winged Crossbill

1. Males 2. Females

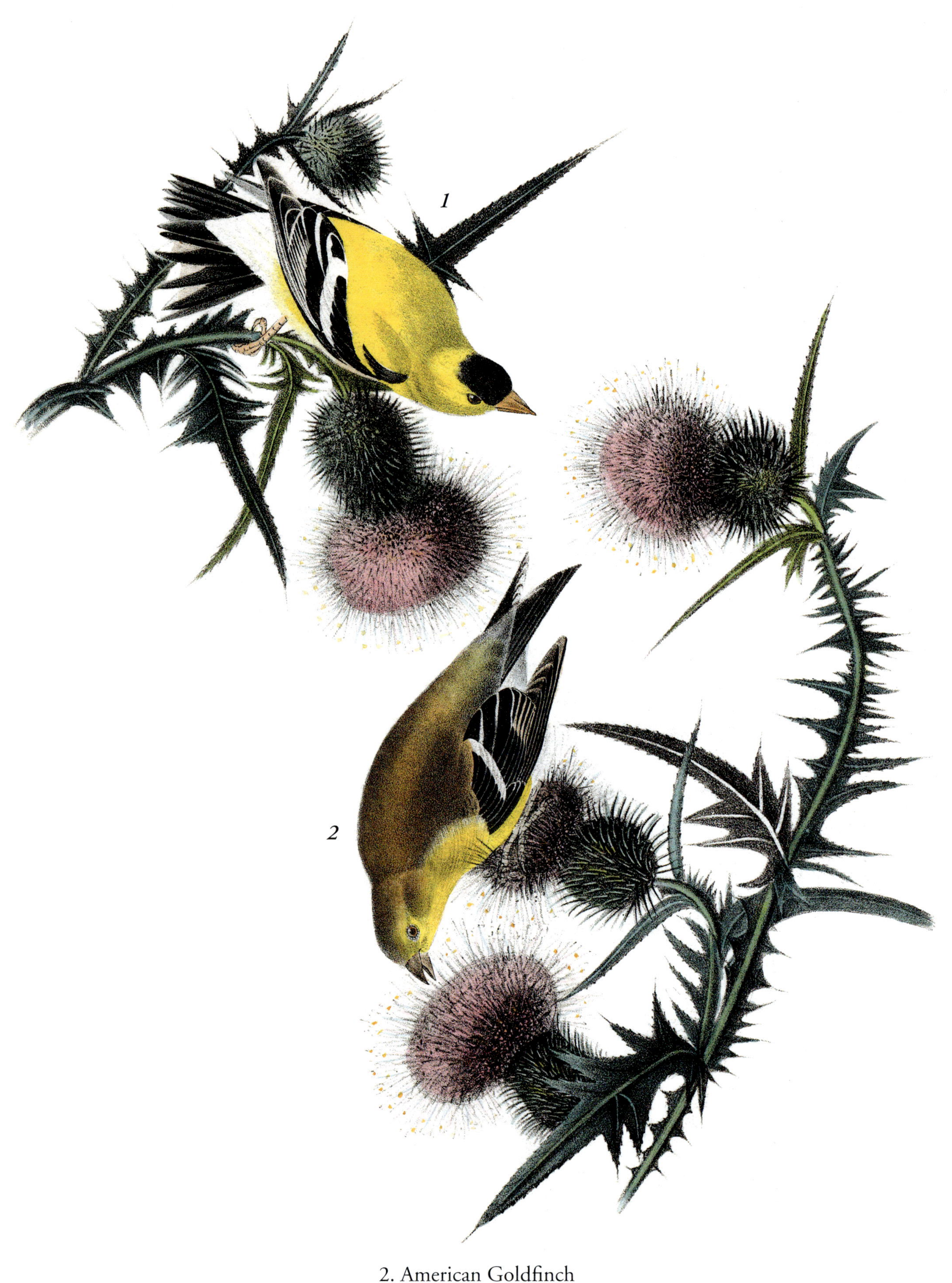

2. American Goldfinch

1. Male 2. Female

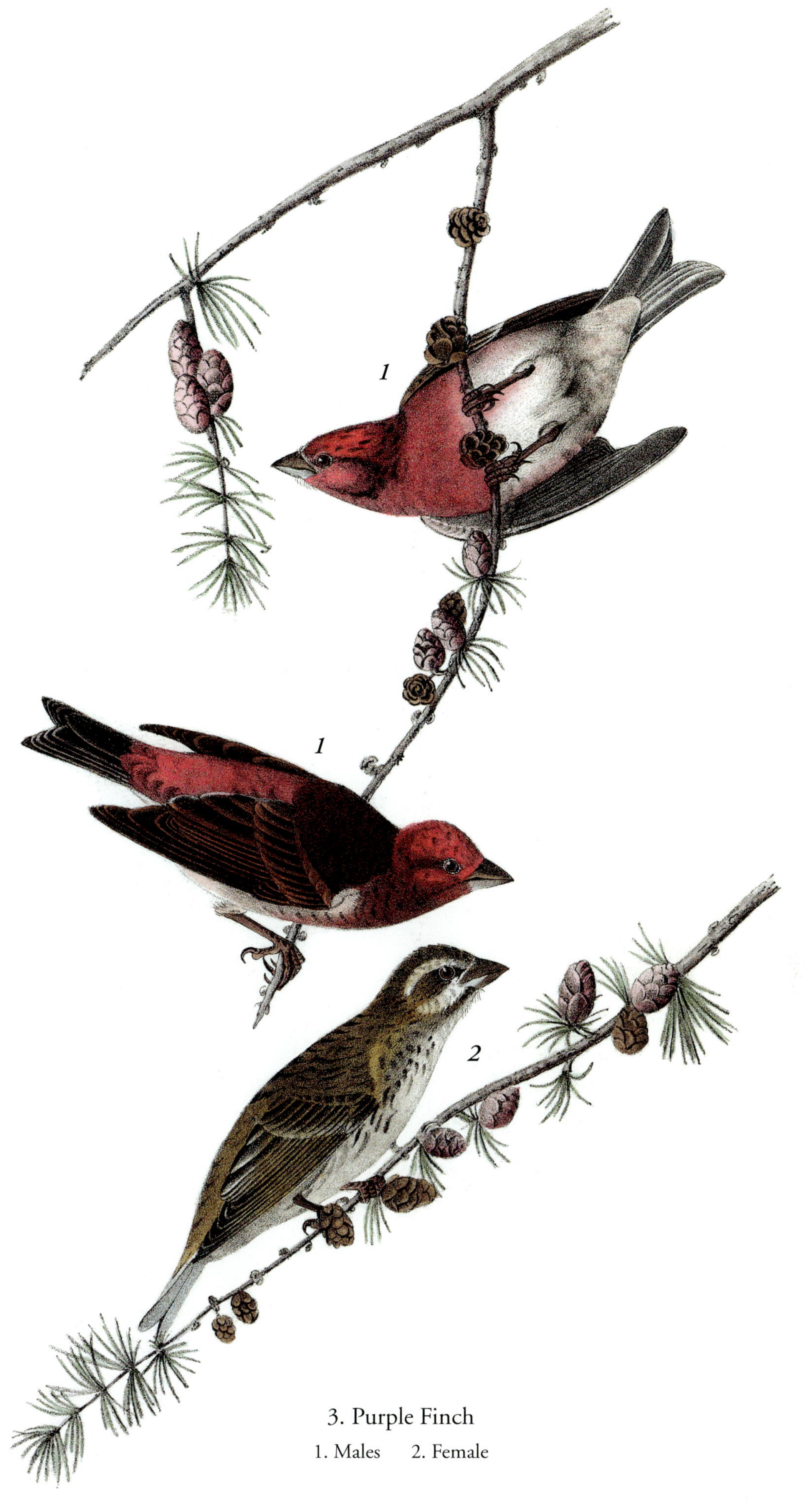

3. Purple Finch

1. Males 2. Female

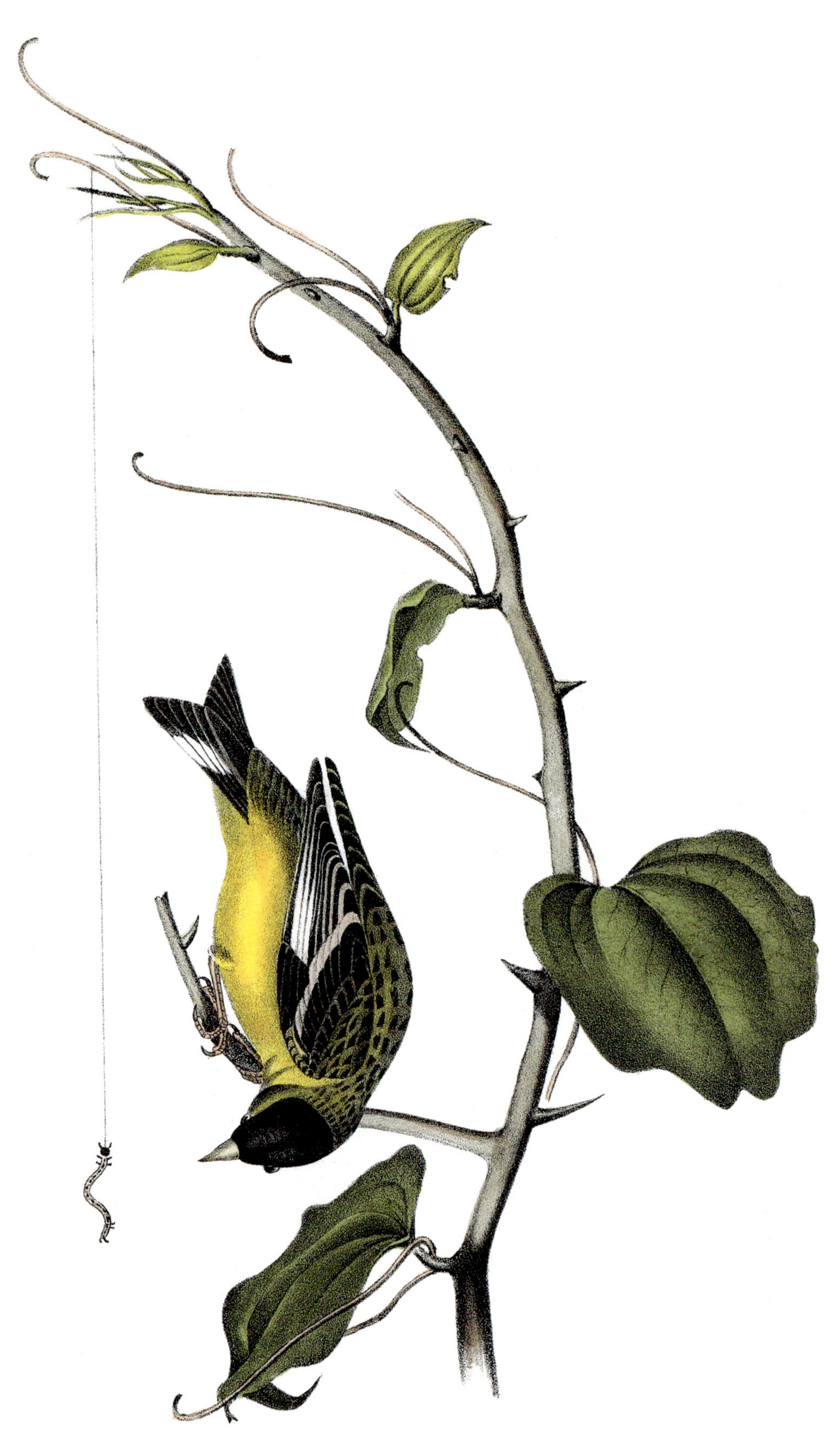

4. Lesser Goldfinch

Male

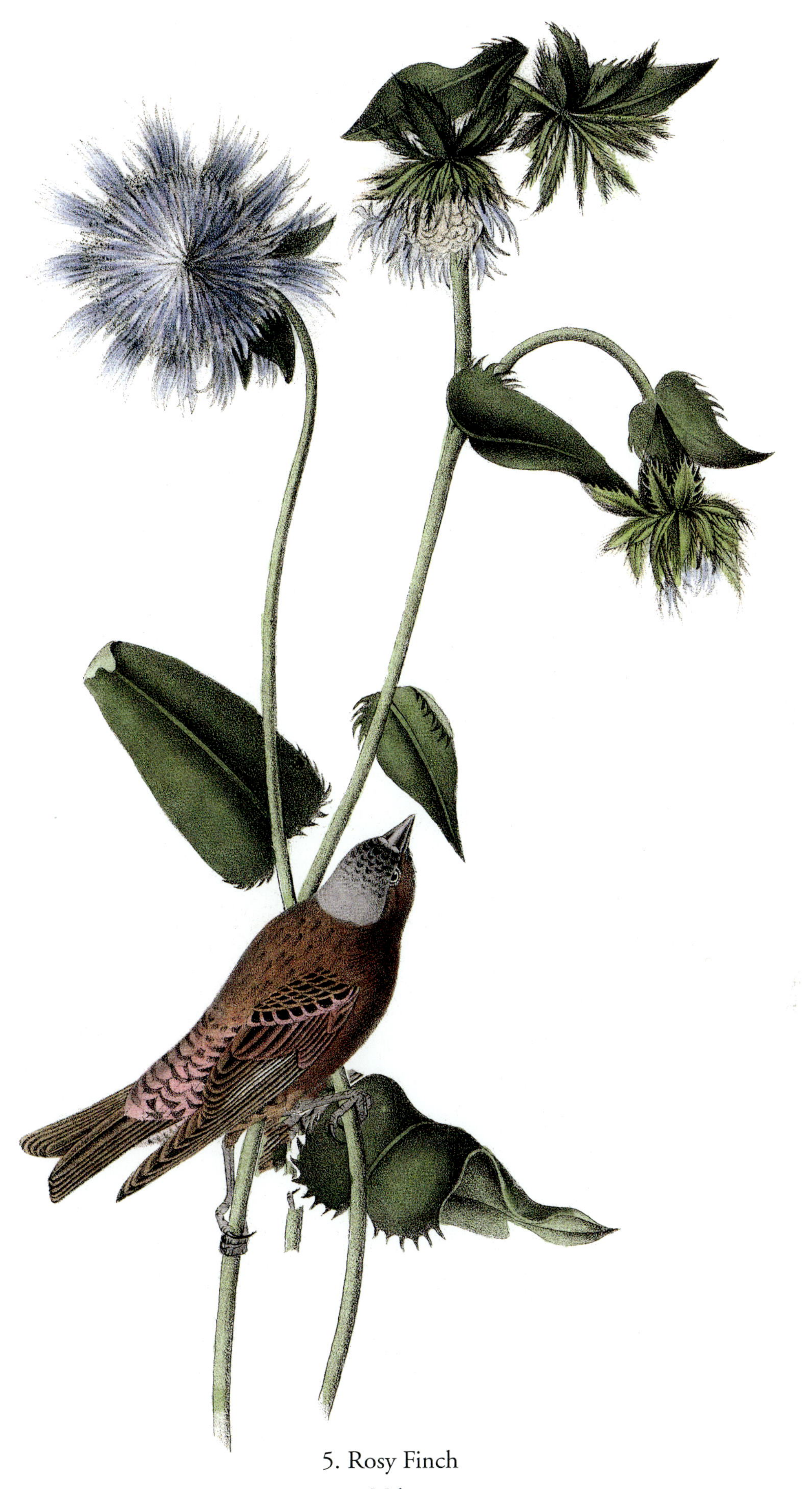

5. Rosy Finch

Male

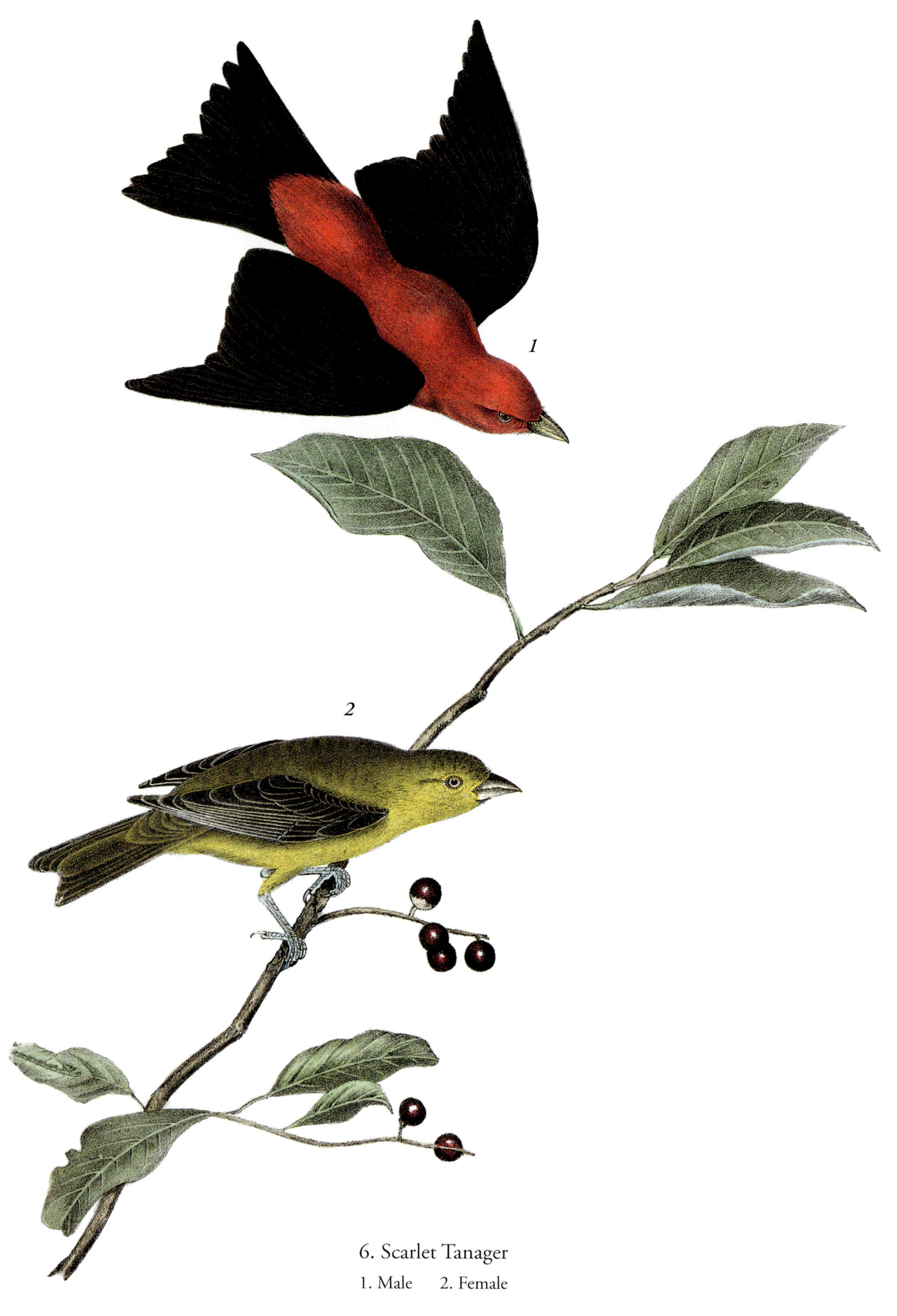

6. Scarlet Tanager

1. Male 2. Female

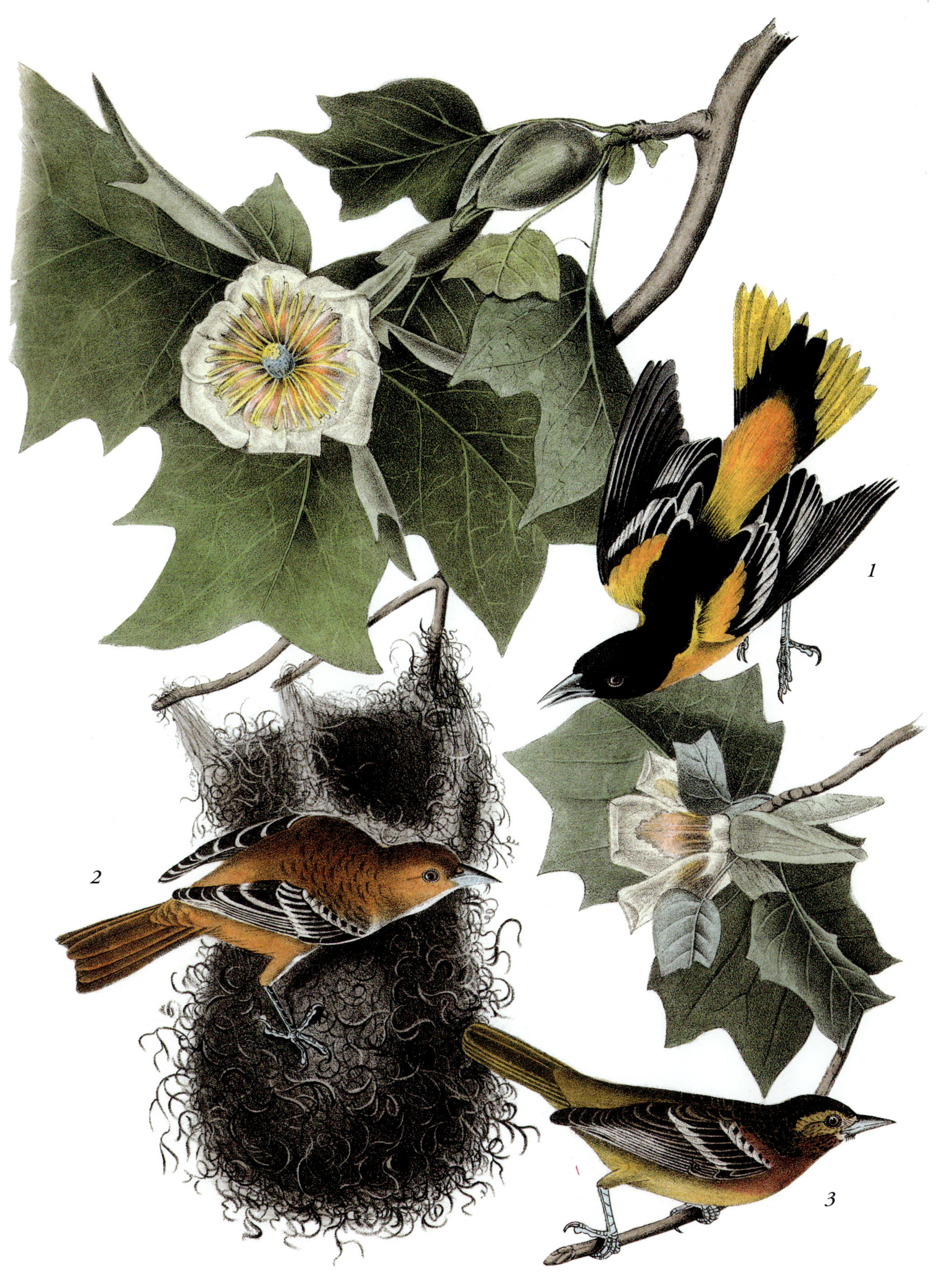

7. Northern Oriole

1. Adult Male 2. Young Male 3. Female

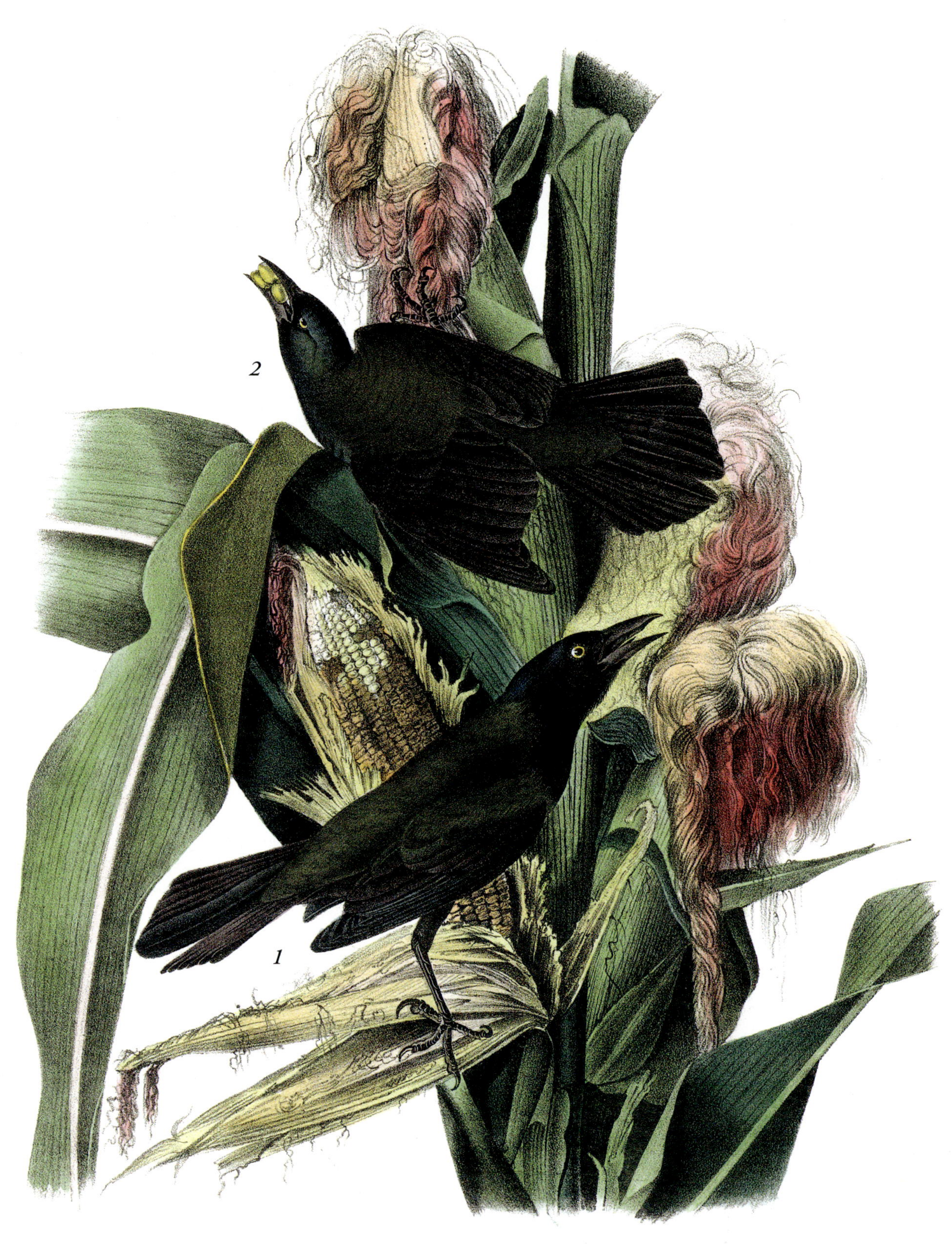

8. Common Grackle

1. Male 2. Female

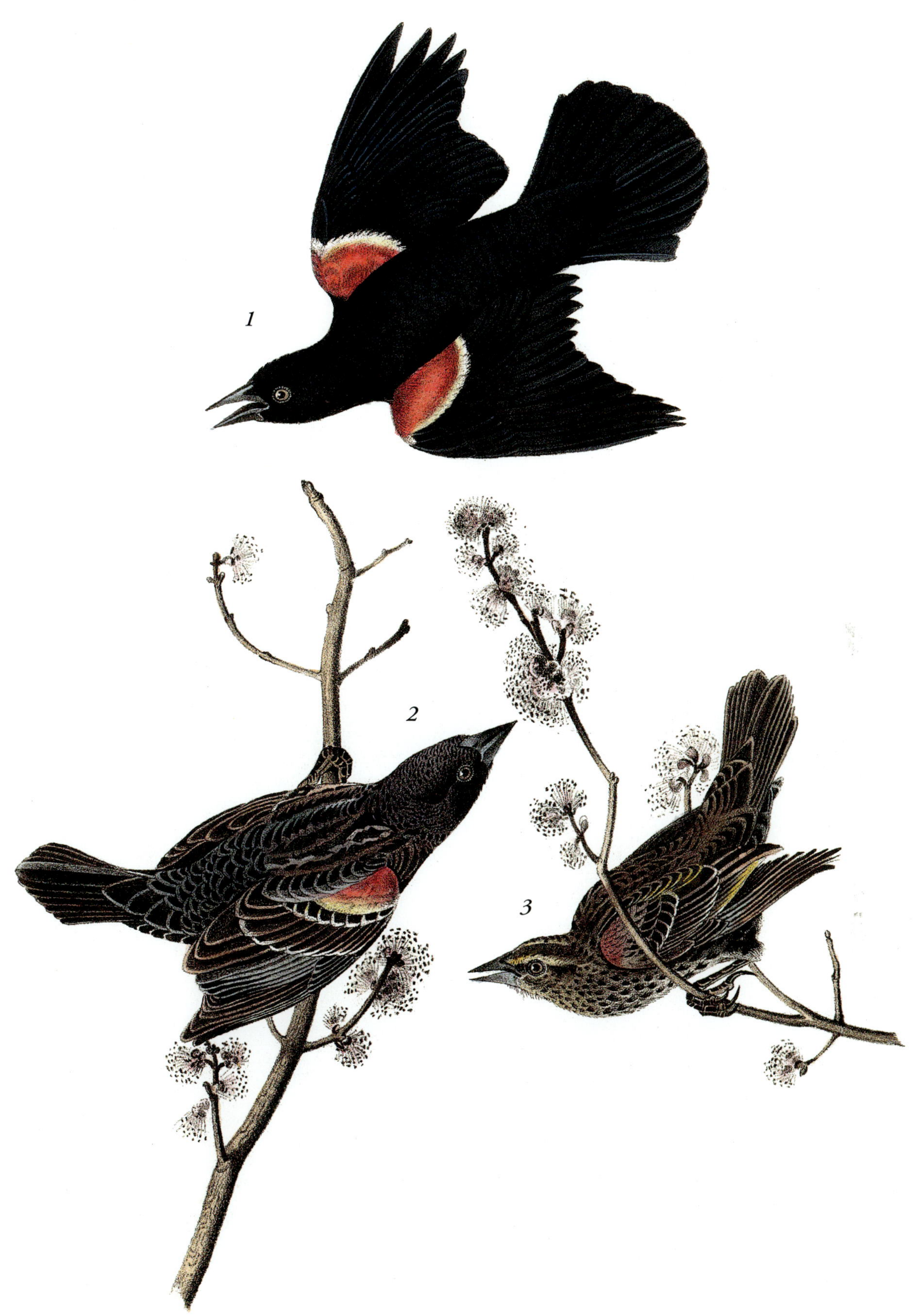

9. Red-Winged Blackbird

1. Adult Male 2. Young Male 3. Female

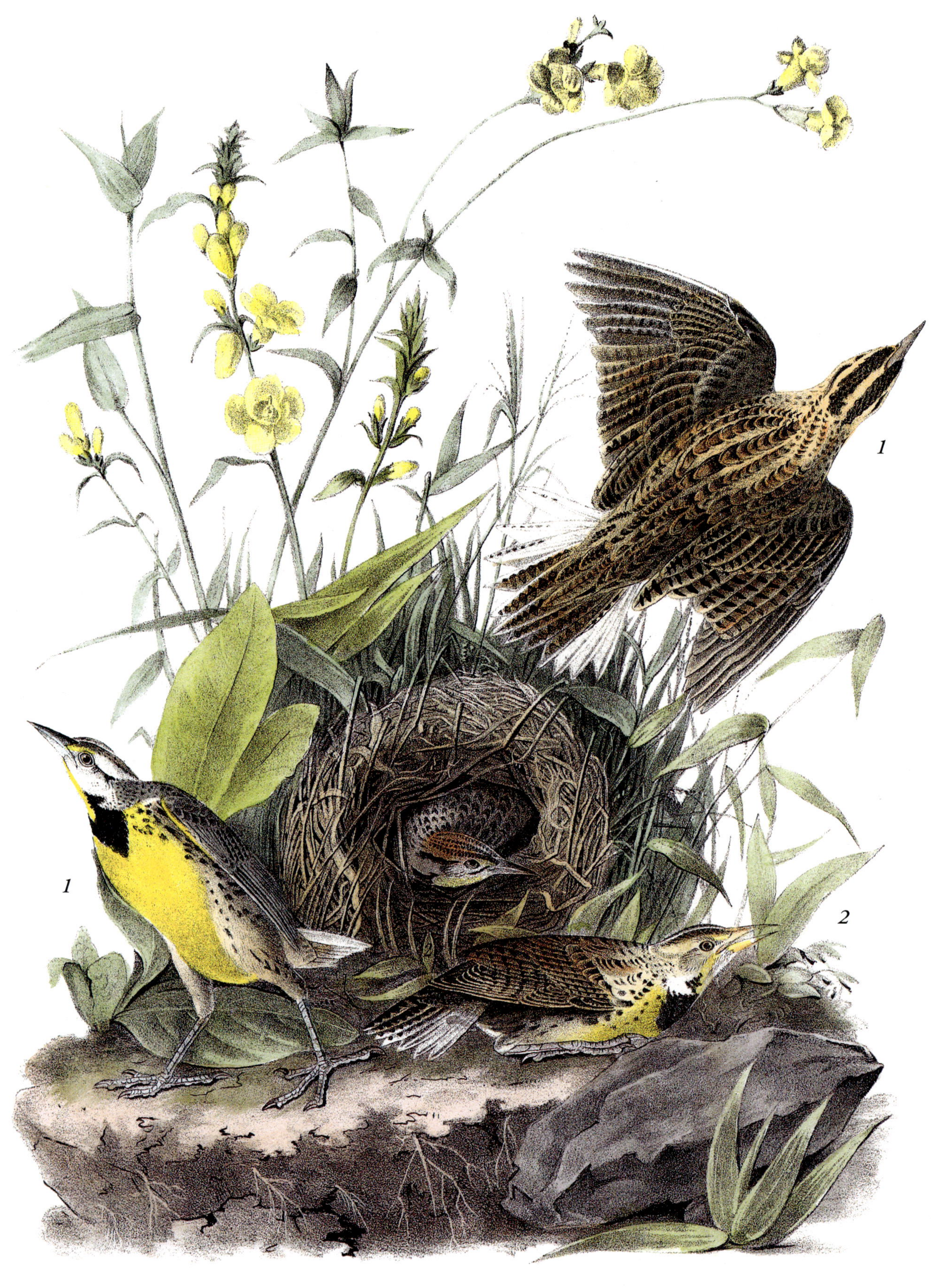

10. Eastern Meadowlark

1. Males 2. Females and Nest

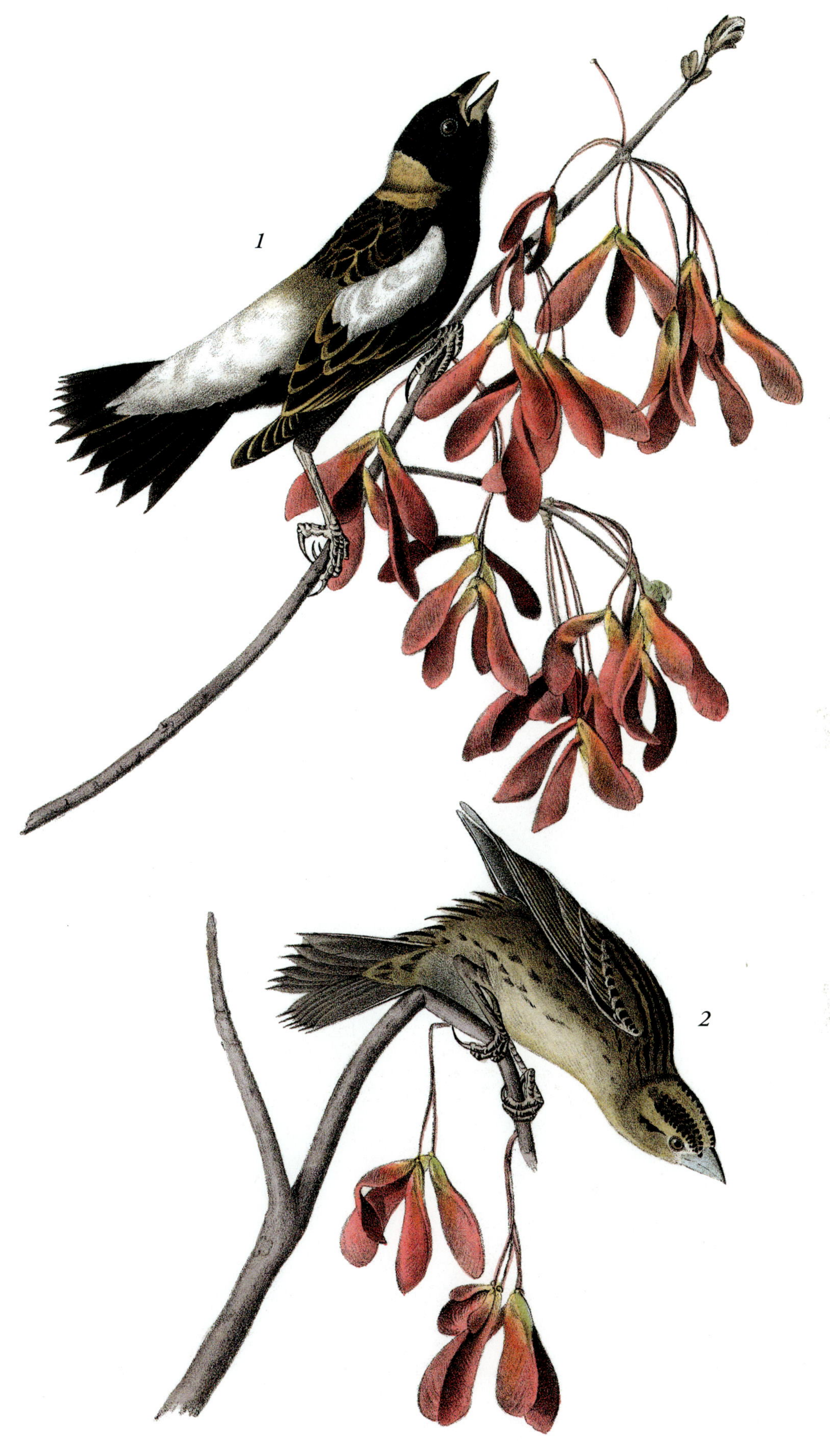

11. Bobolink

1. Male 2. Female

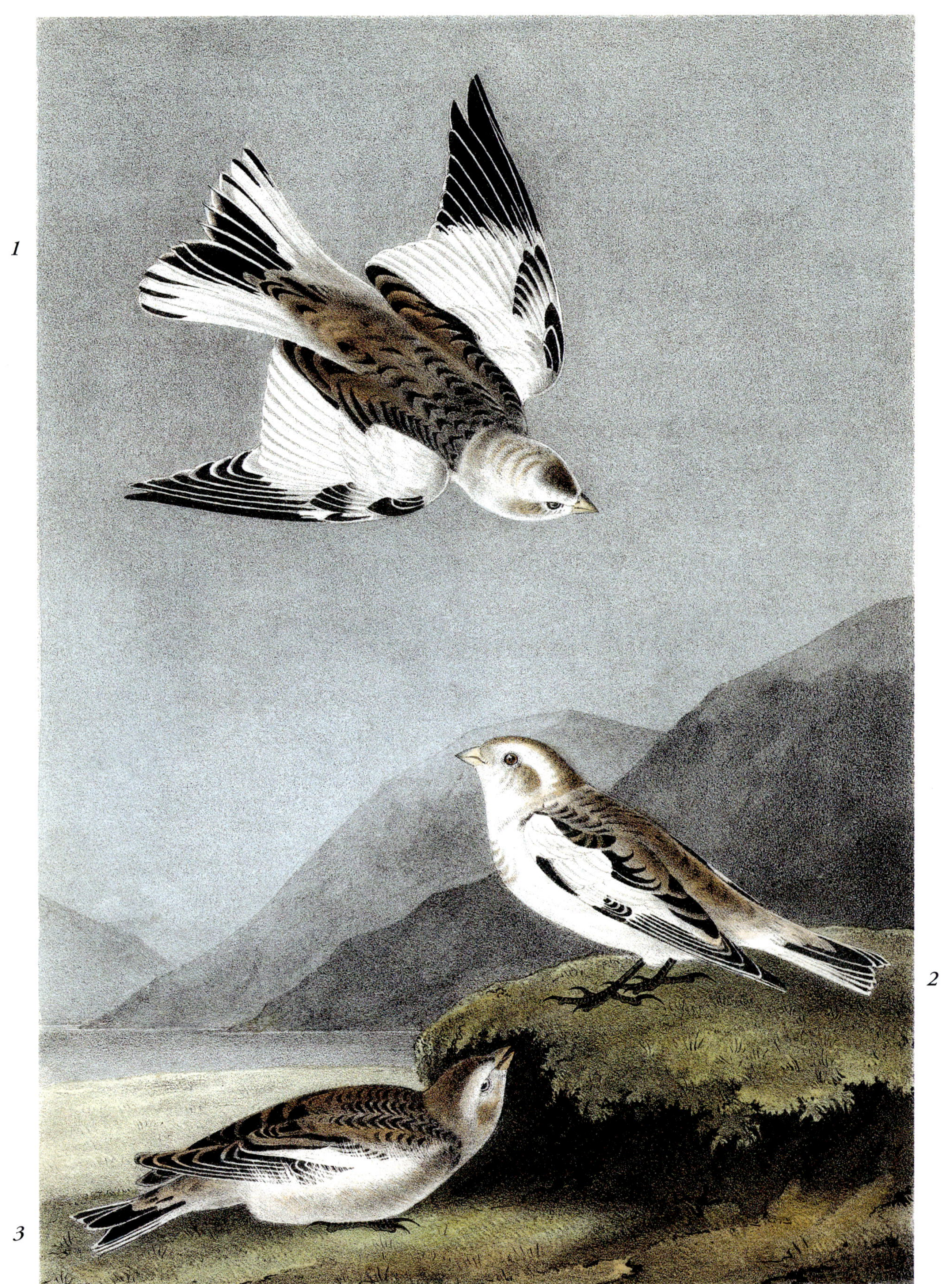

12. Snow Bunting

1, 2. Adult 3. Young

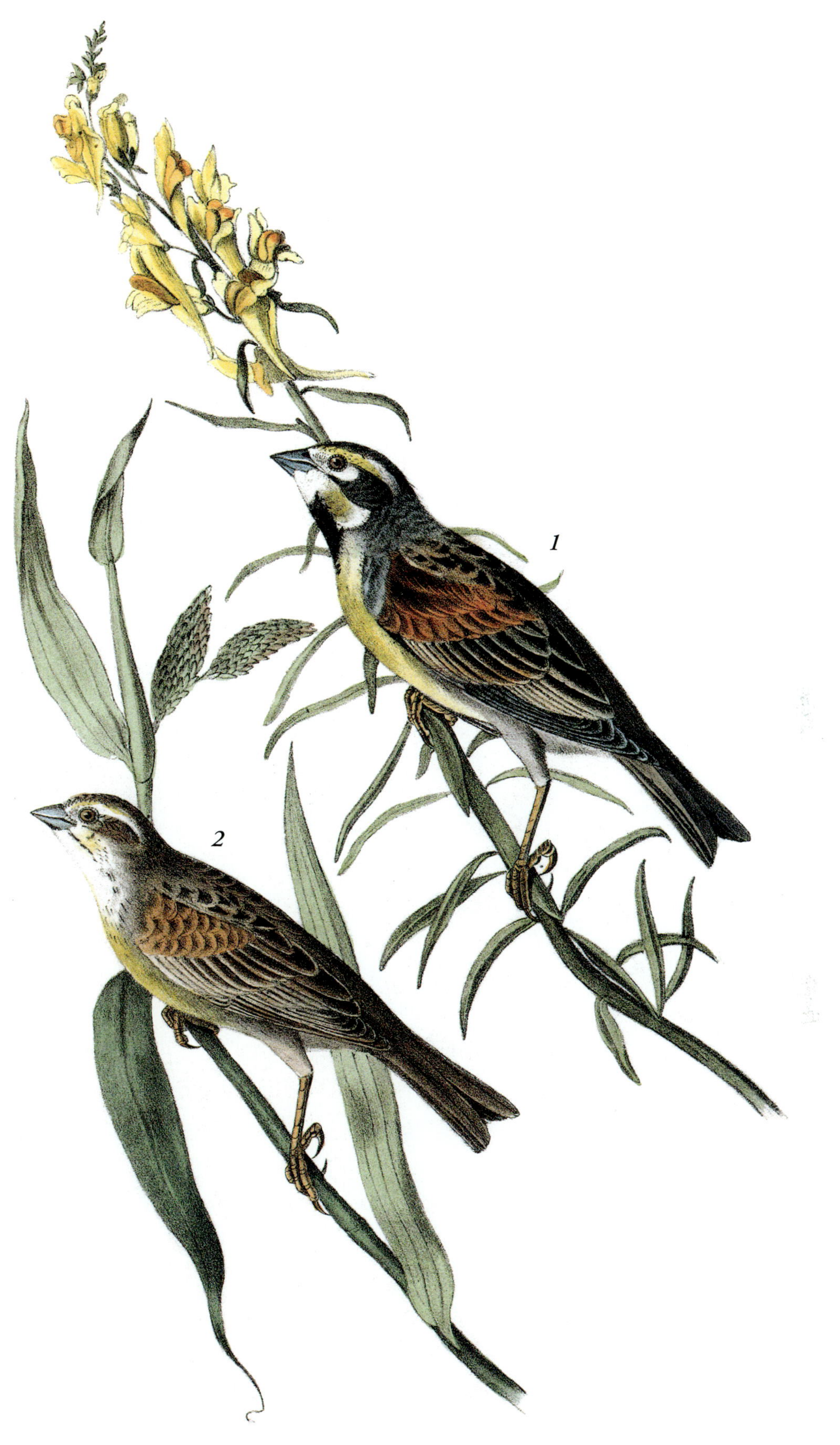

13. Dickcissel

1. Male 2. Female

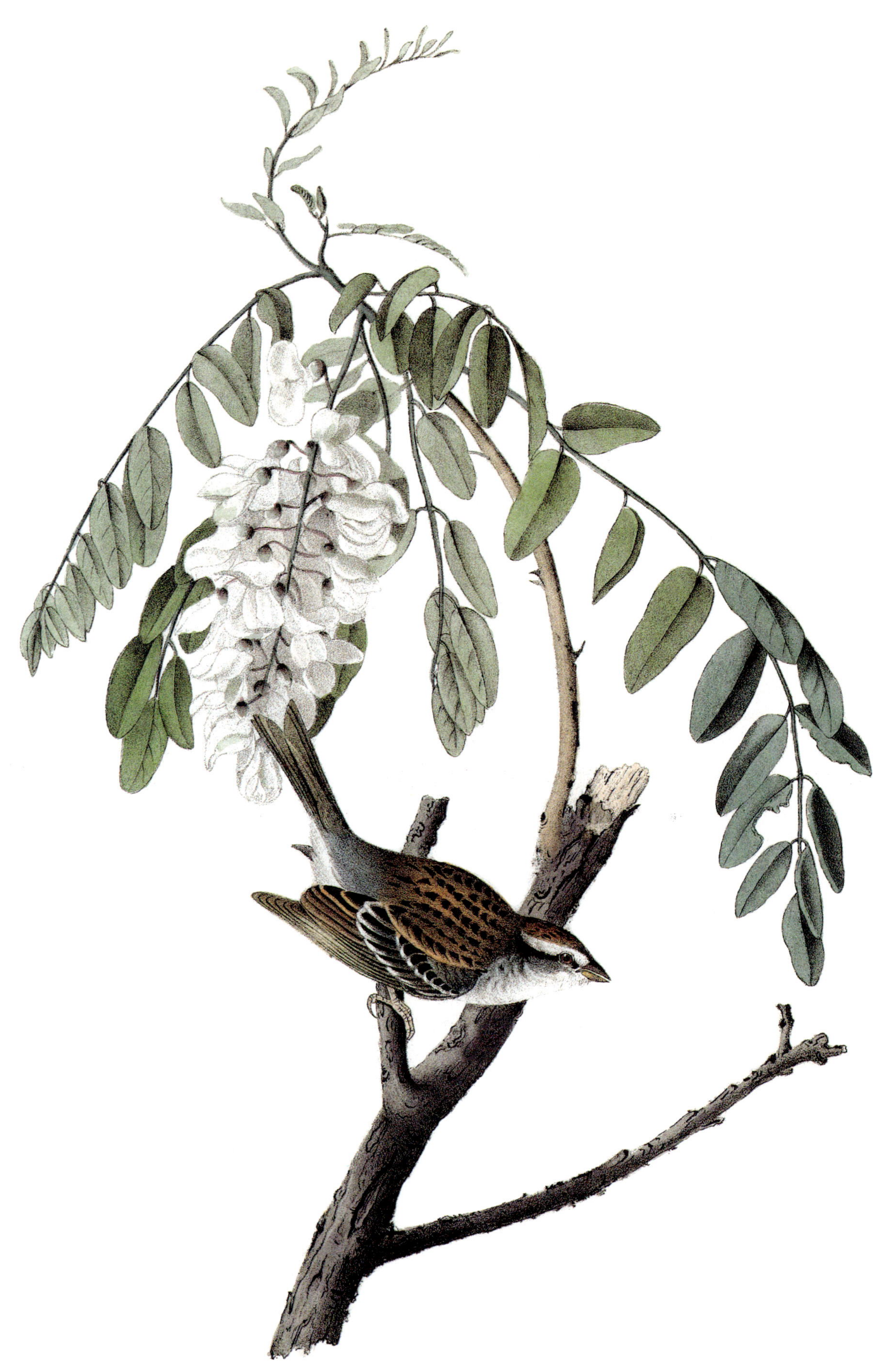

14. Chipping Sparrow
Male

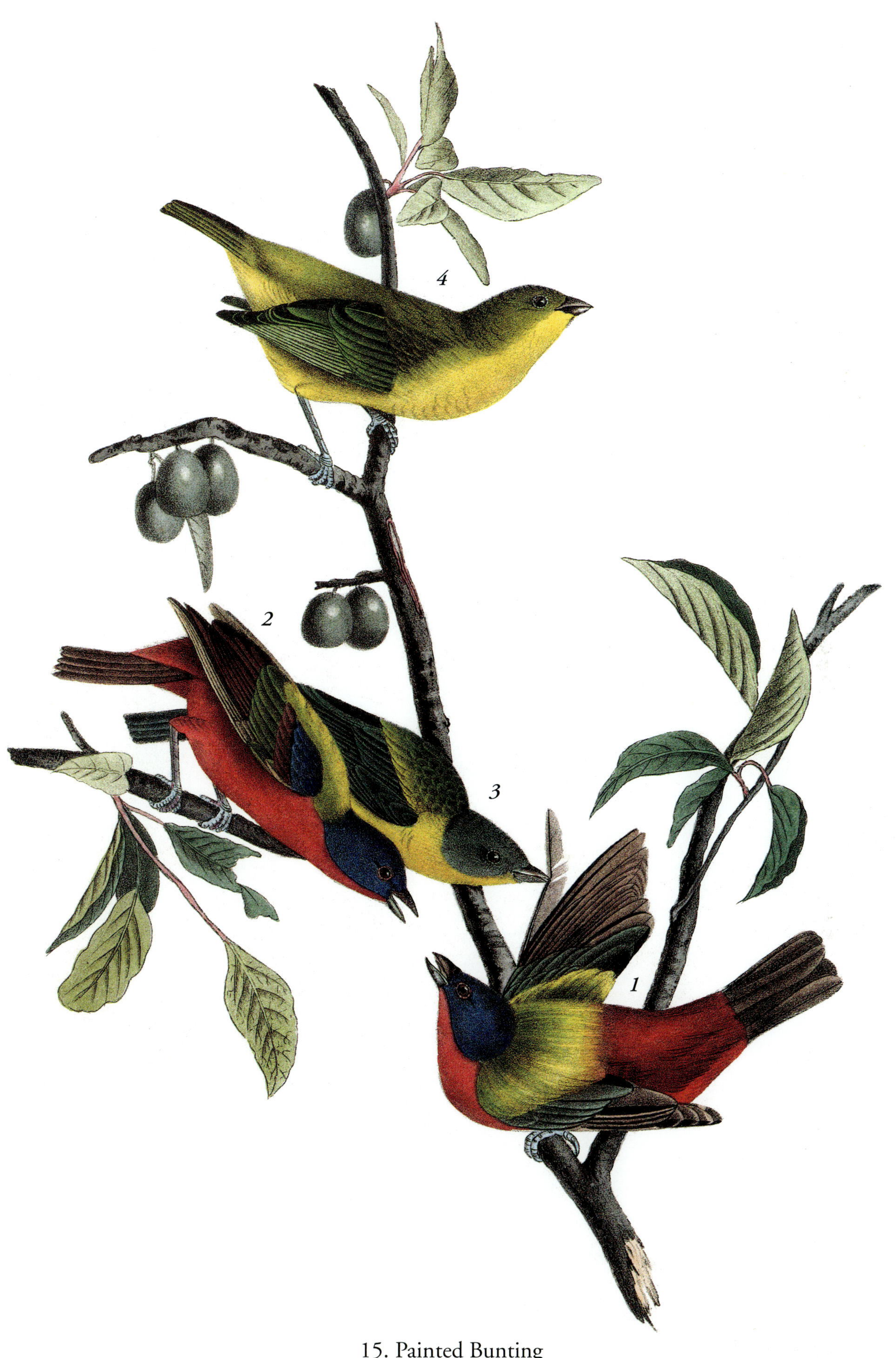

15. Painted Bunting

1, 2, 3. Males, Different States of Plumage 4. Female

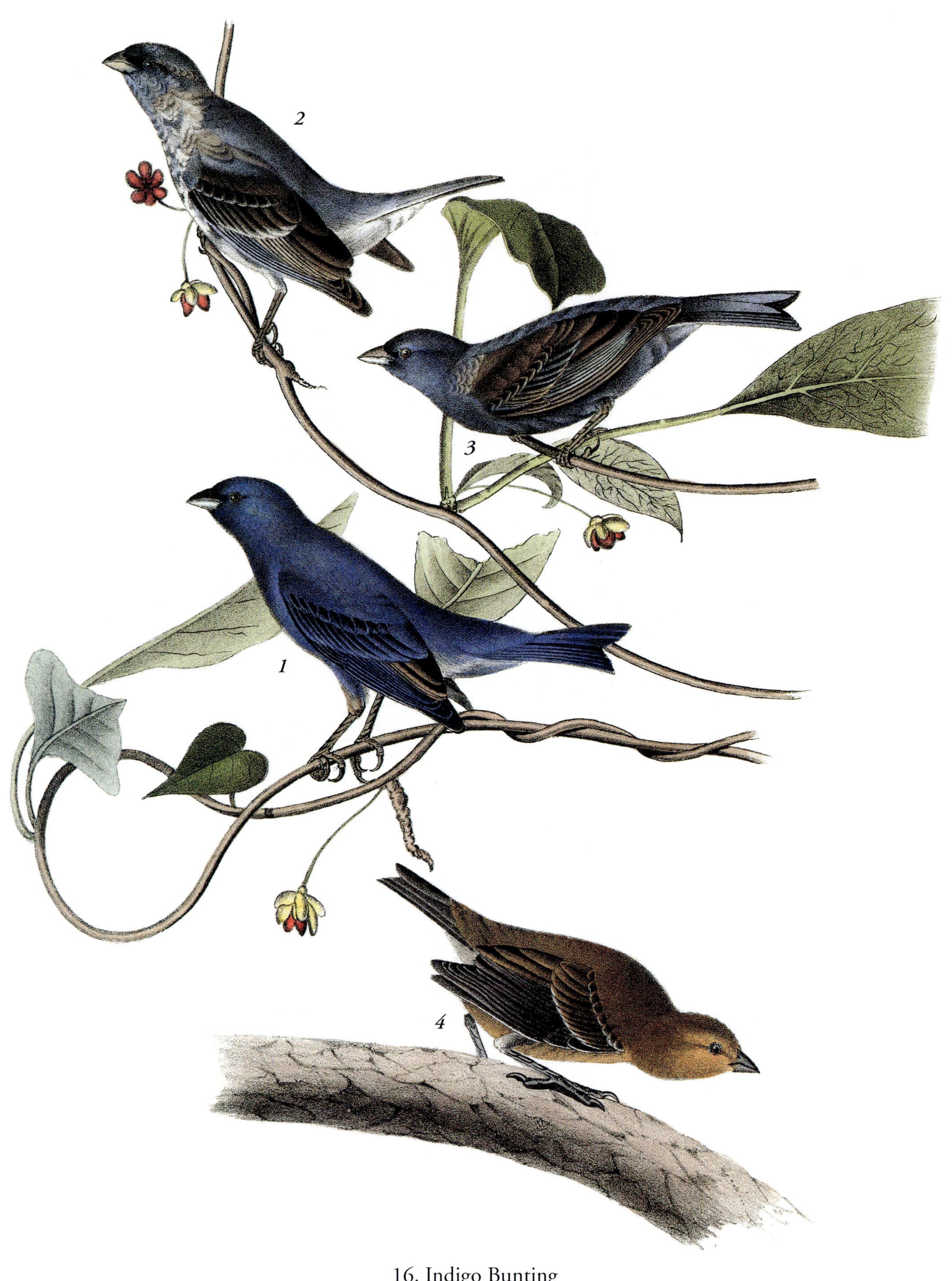

16. Indigo Bunting

1, 2, 3. Males, Different States of Plumage   4. Female

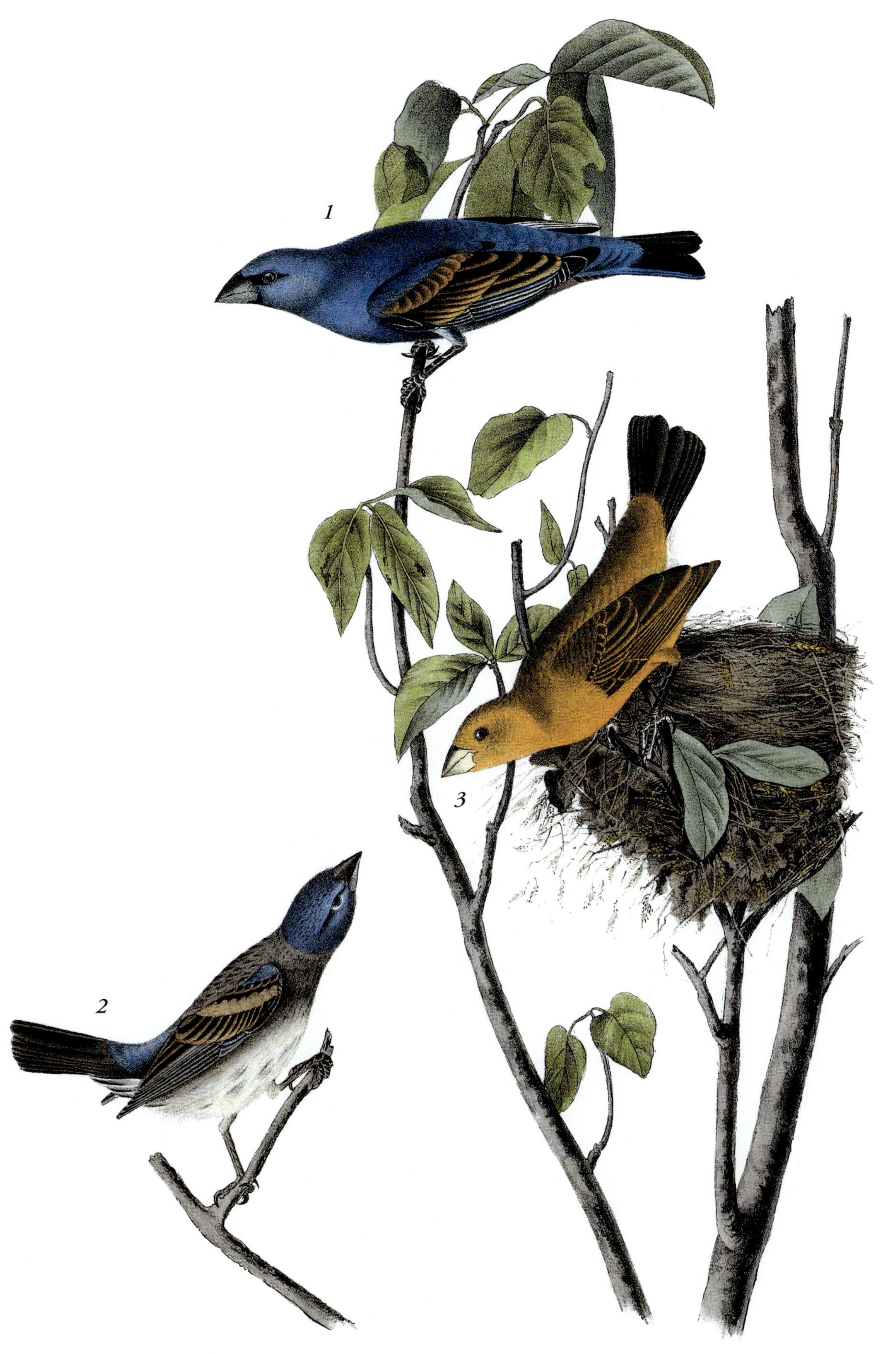

17. Blue Grosbeak

1. Male 2. Female 3. Young

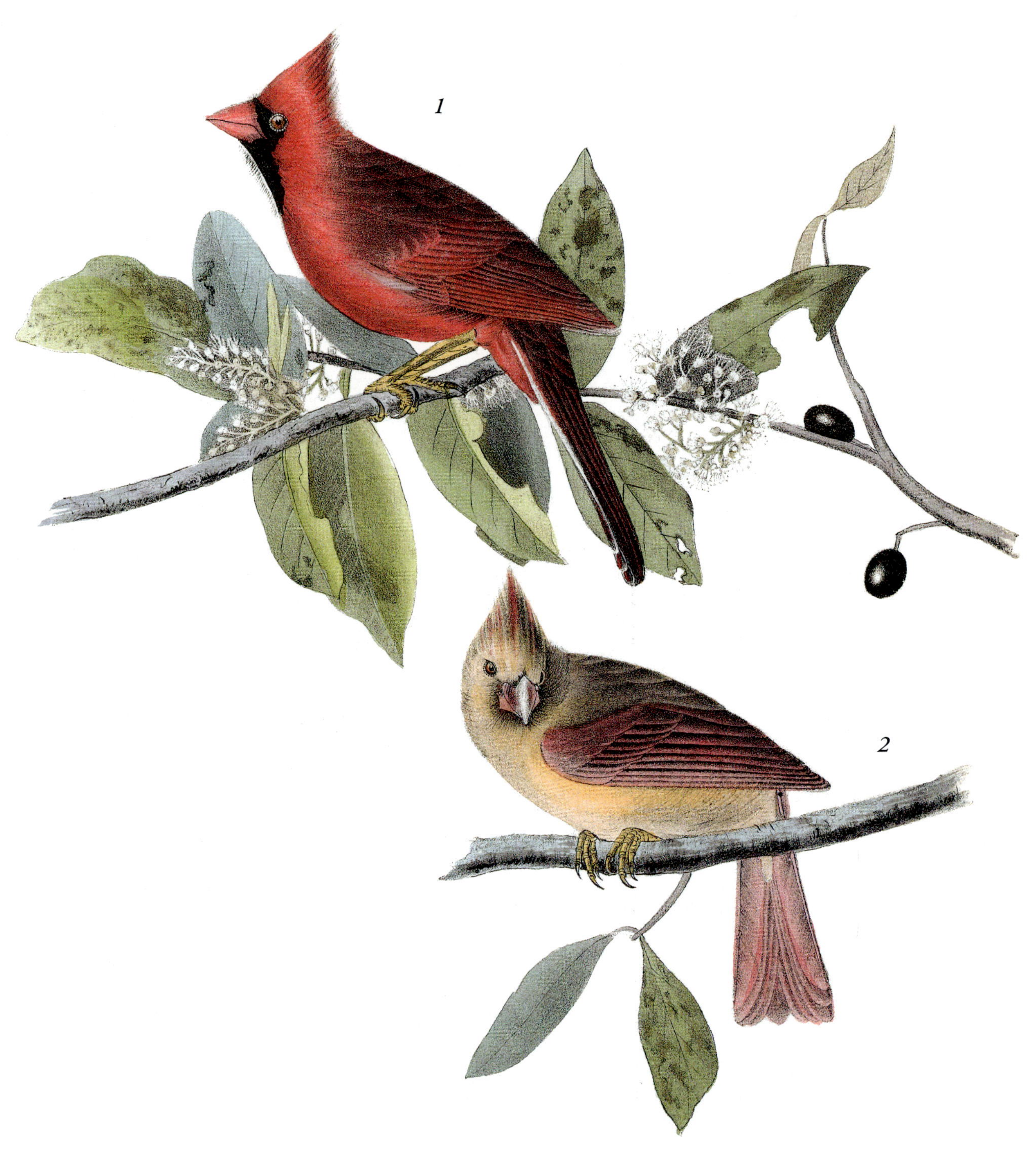

18. Northern Cardinal

1. Male 2. Female

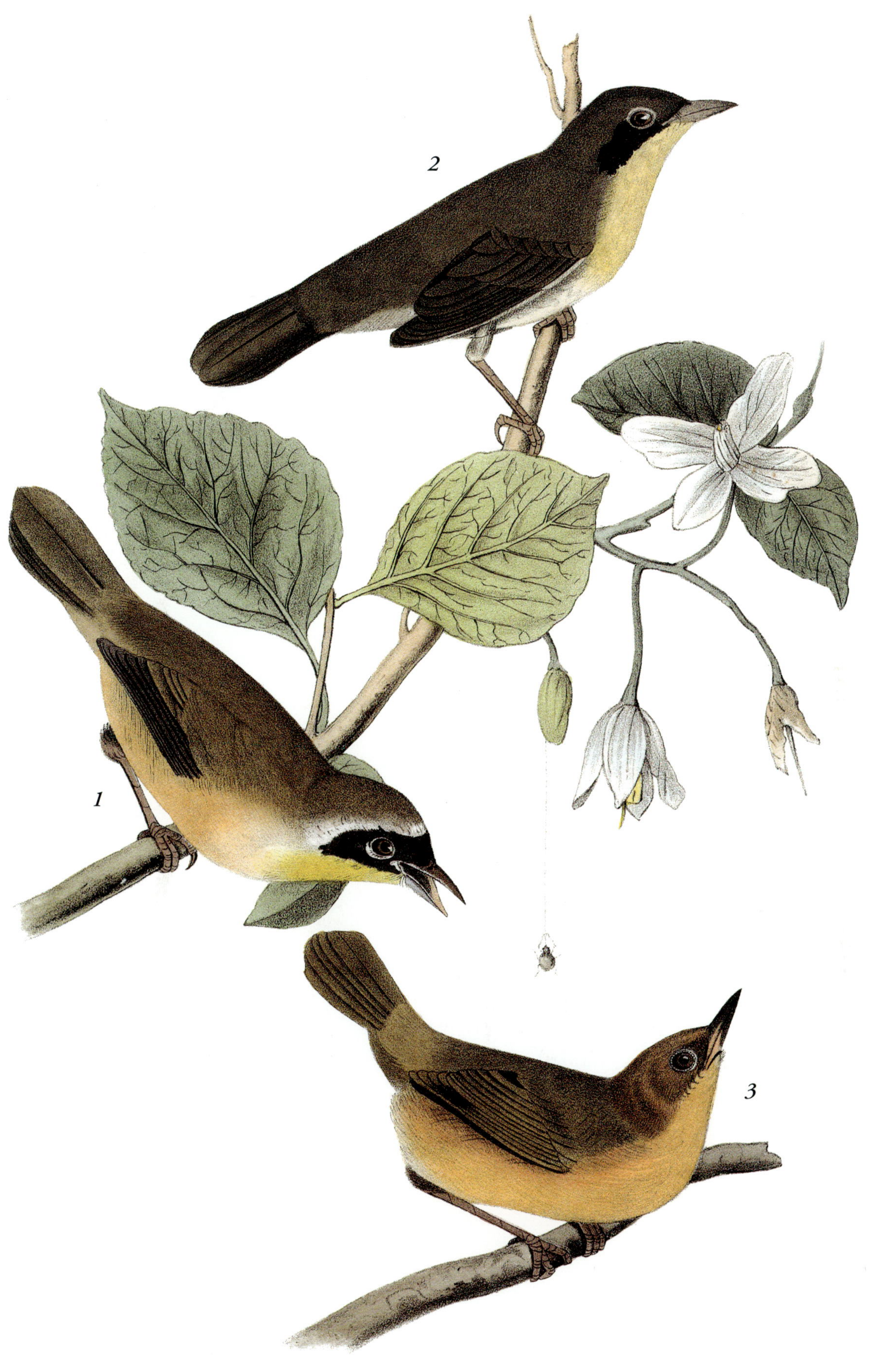

19. Common Yellowthroat

1. Adult Male   2. Young Male   3. Female

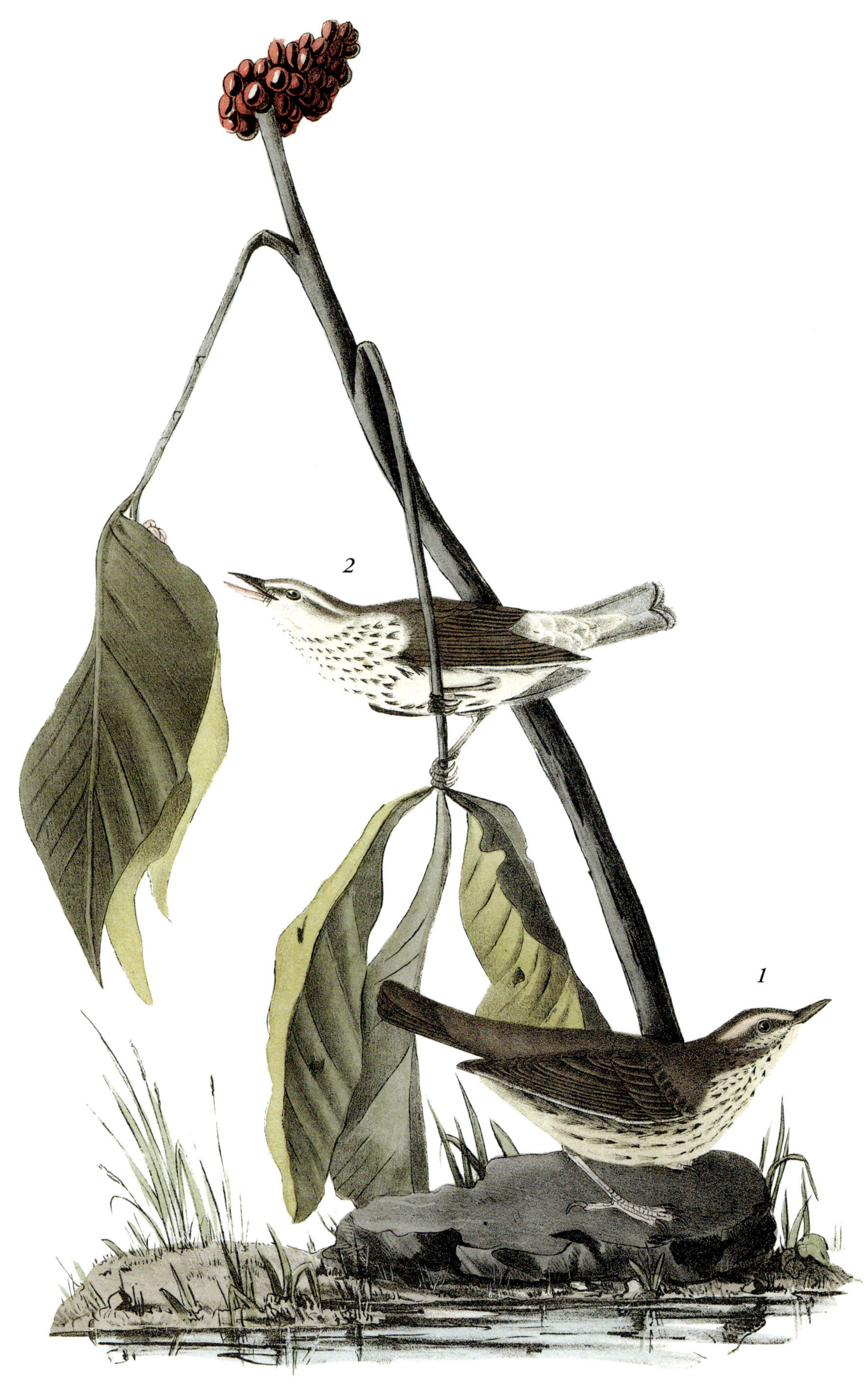

20. Northern Waterthrush

1. Male 2. Female

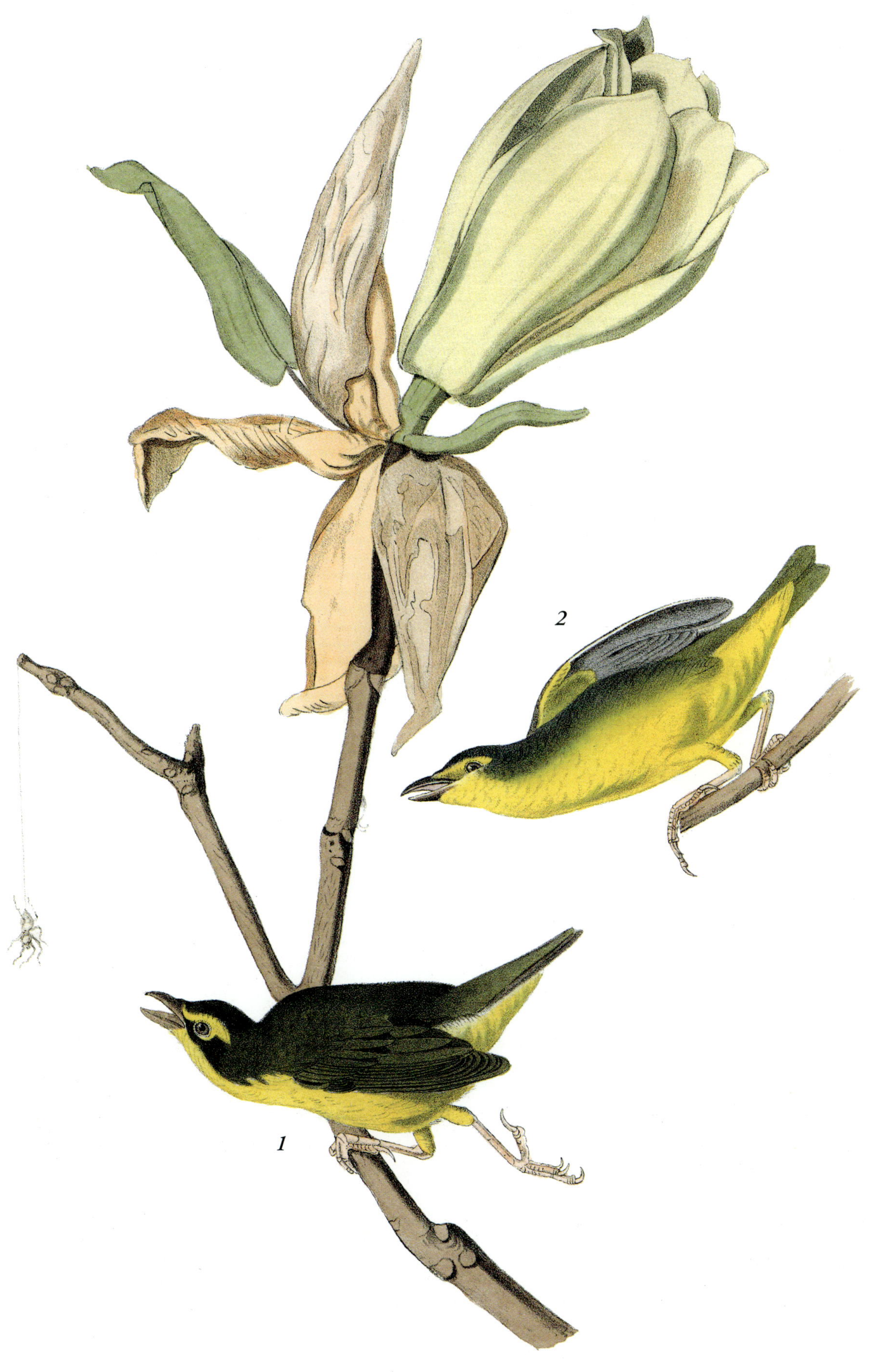

21. Kentucky Warbler

1. Male   2. Female

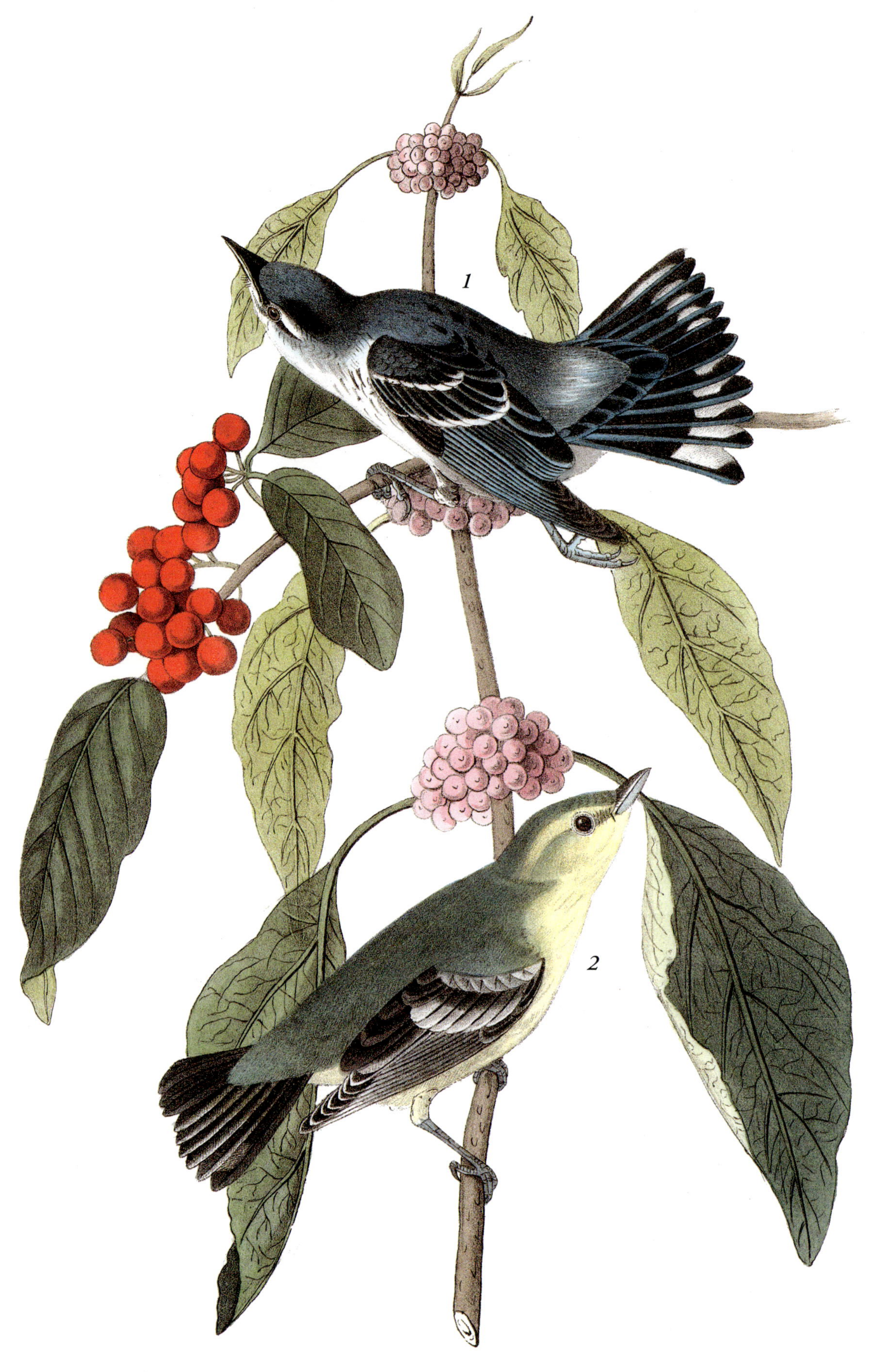

22. Cerulean Warbler

1. Old Male 2. Young Male

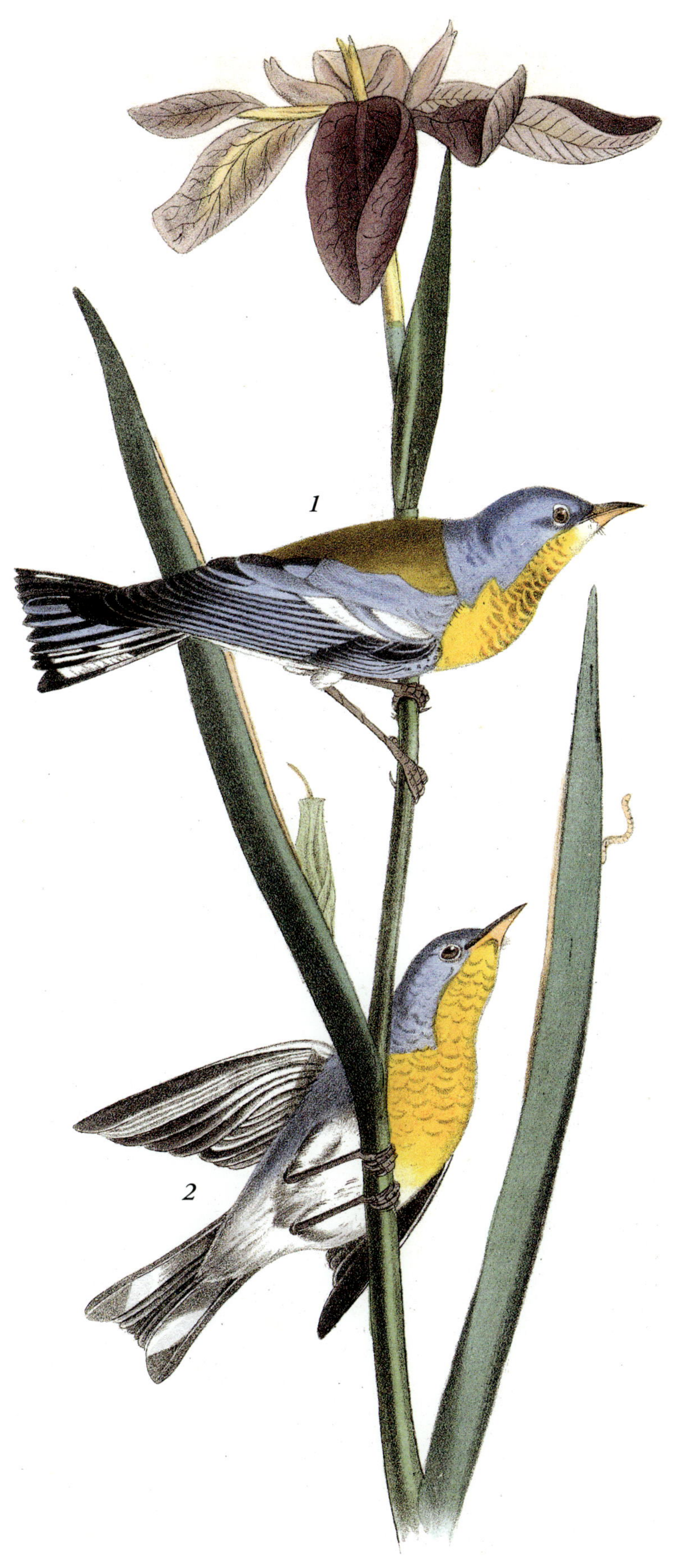

23. Northern Parula

1. Male 2. Female

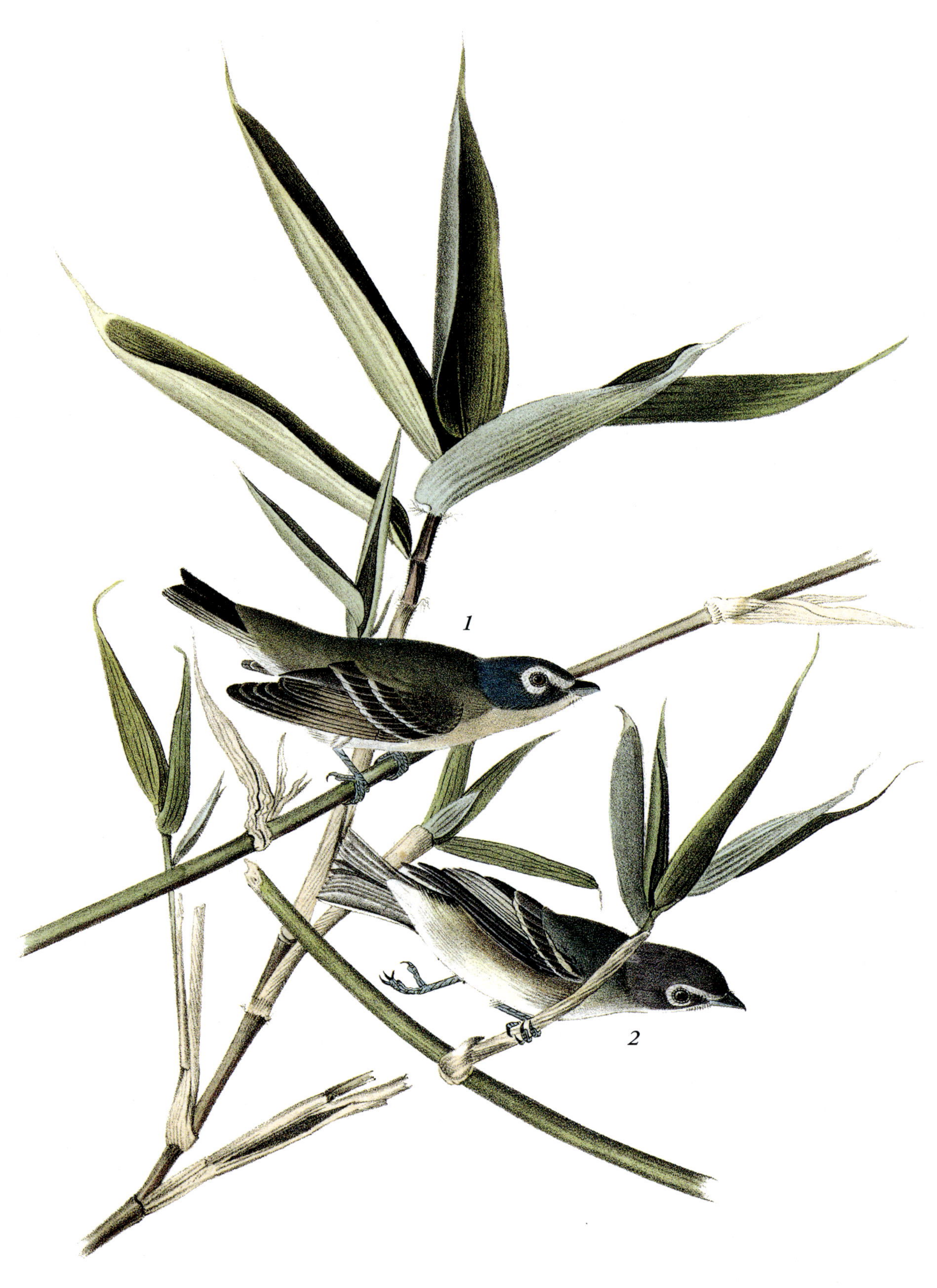

24. Solitary Vireo

1. Male 2. Female

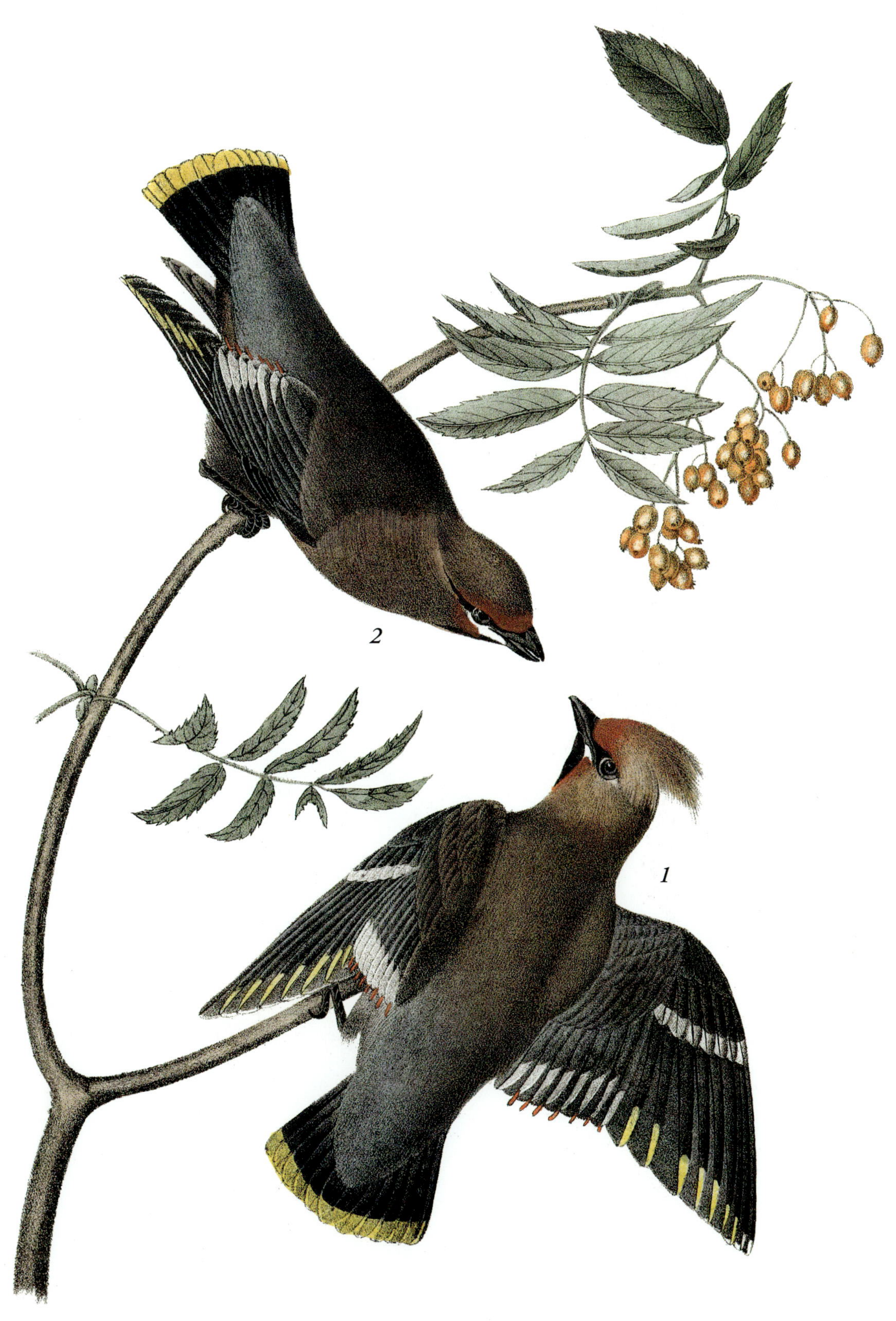

25. Bohemian Waxwing

1. Male 2. Female

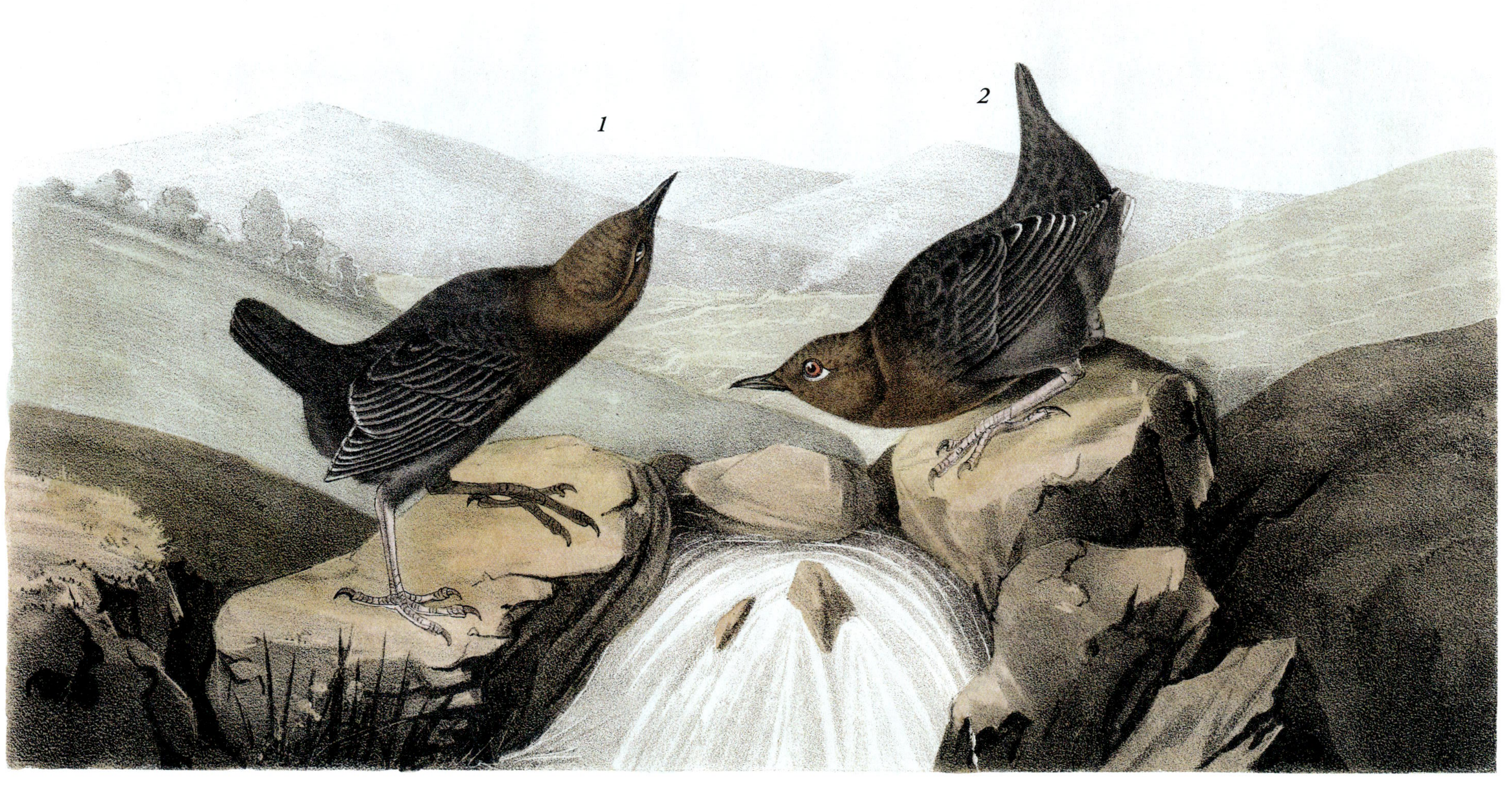

26. American Dipper

1. Male 2. Female

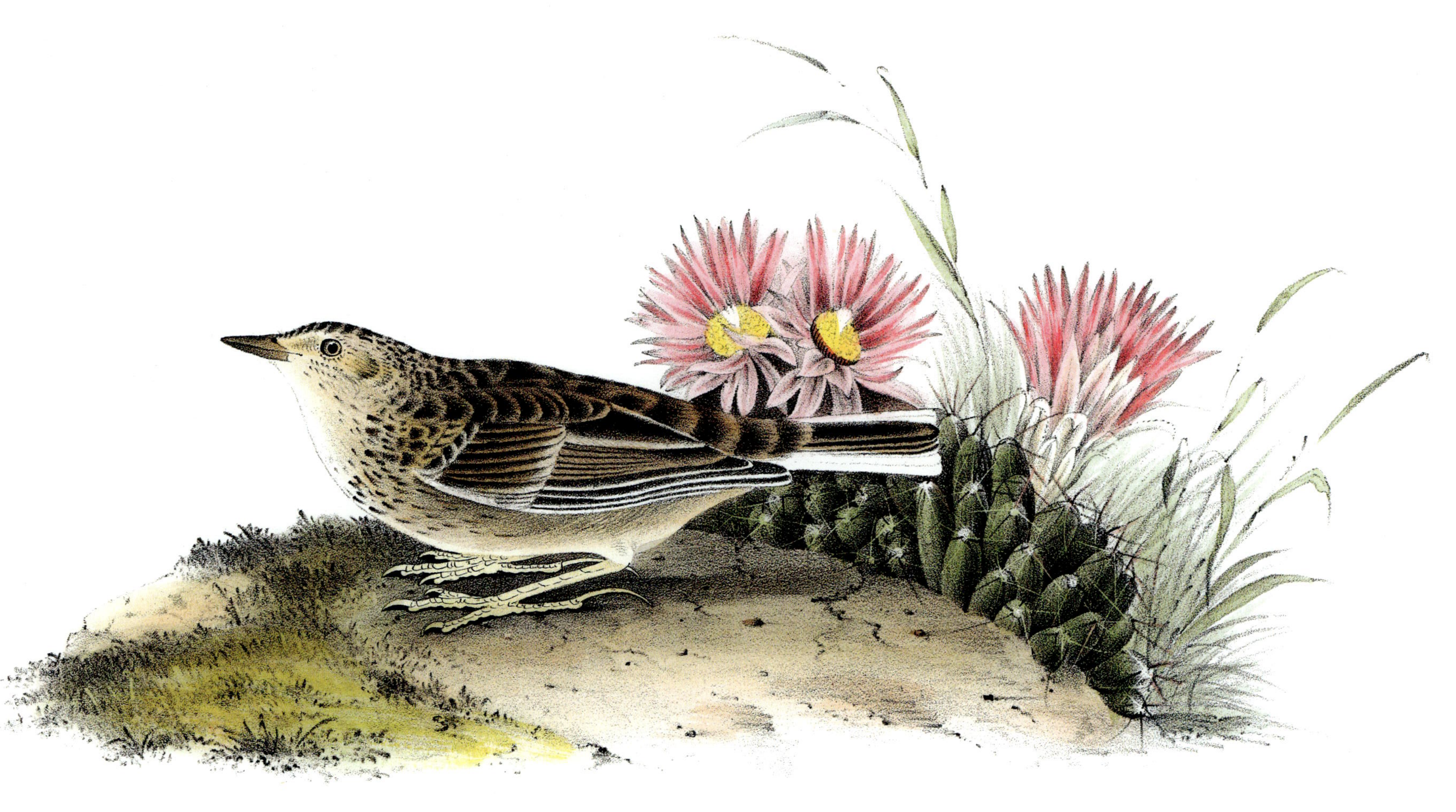

27. Sprague's Pipit
Male

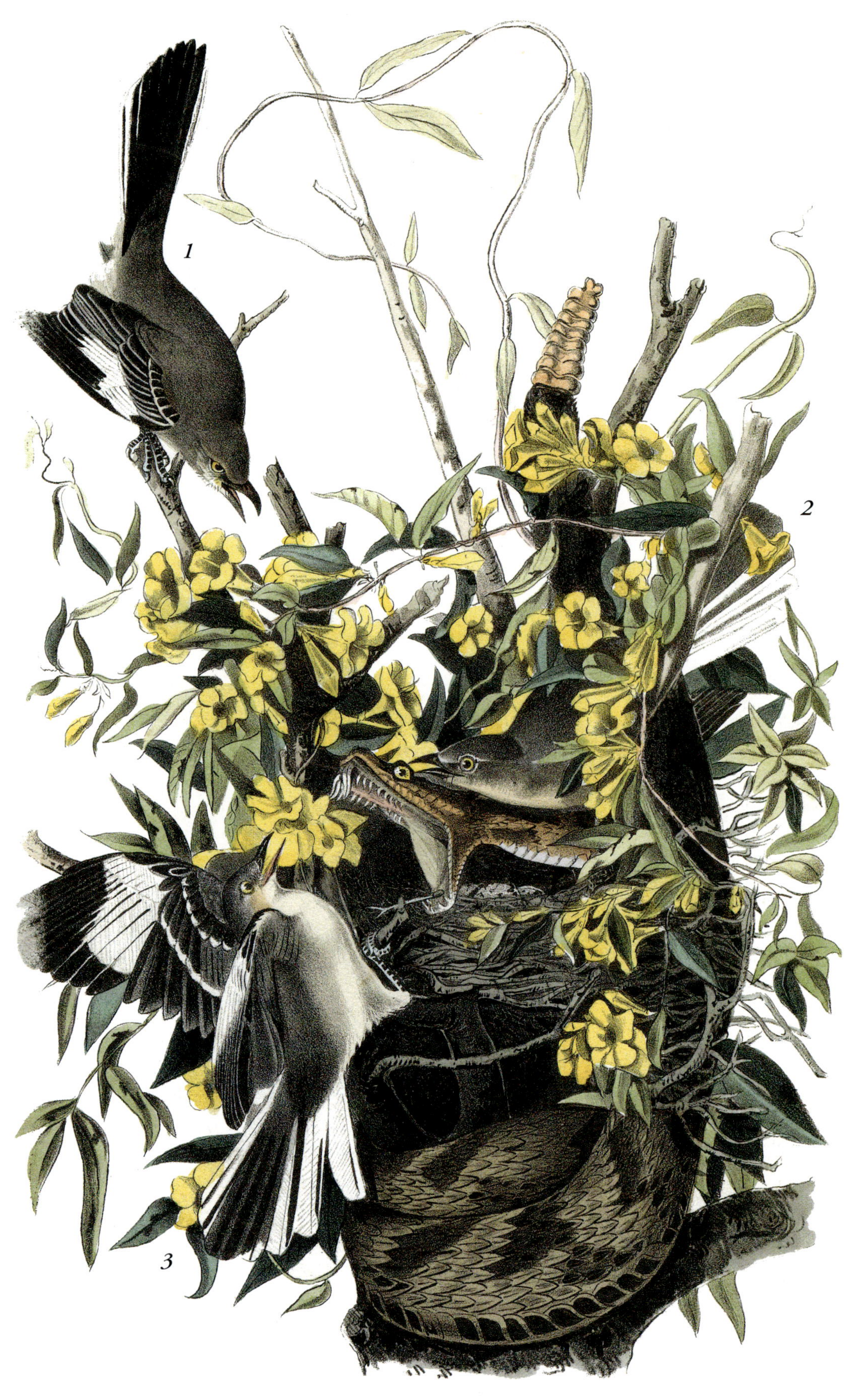

28. Northern Mockingbird

1, 2. Males 3. Female

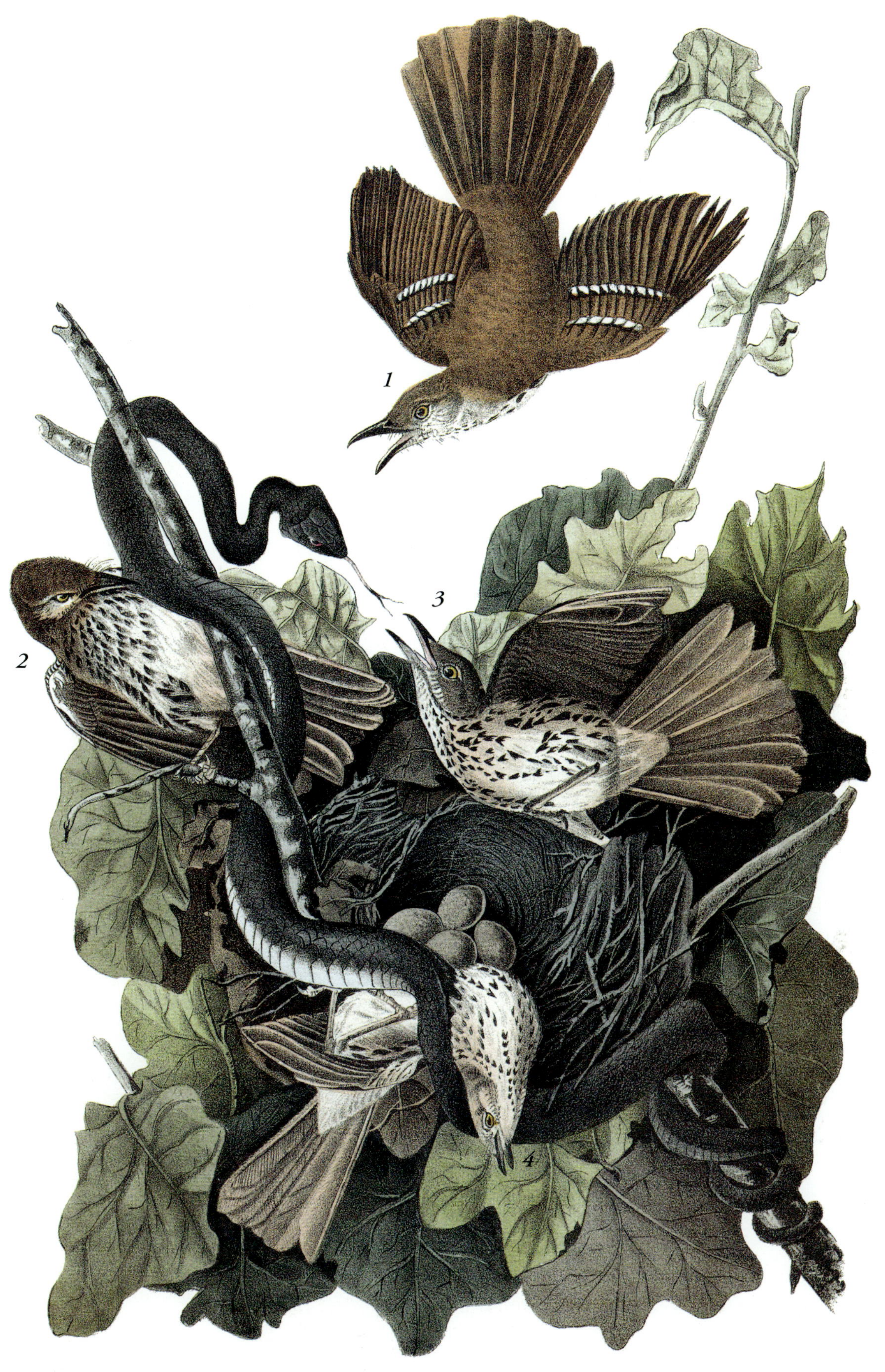

29. Brown Thrasher

1, 2, 3. Males 4. Female

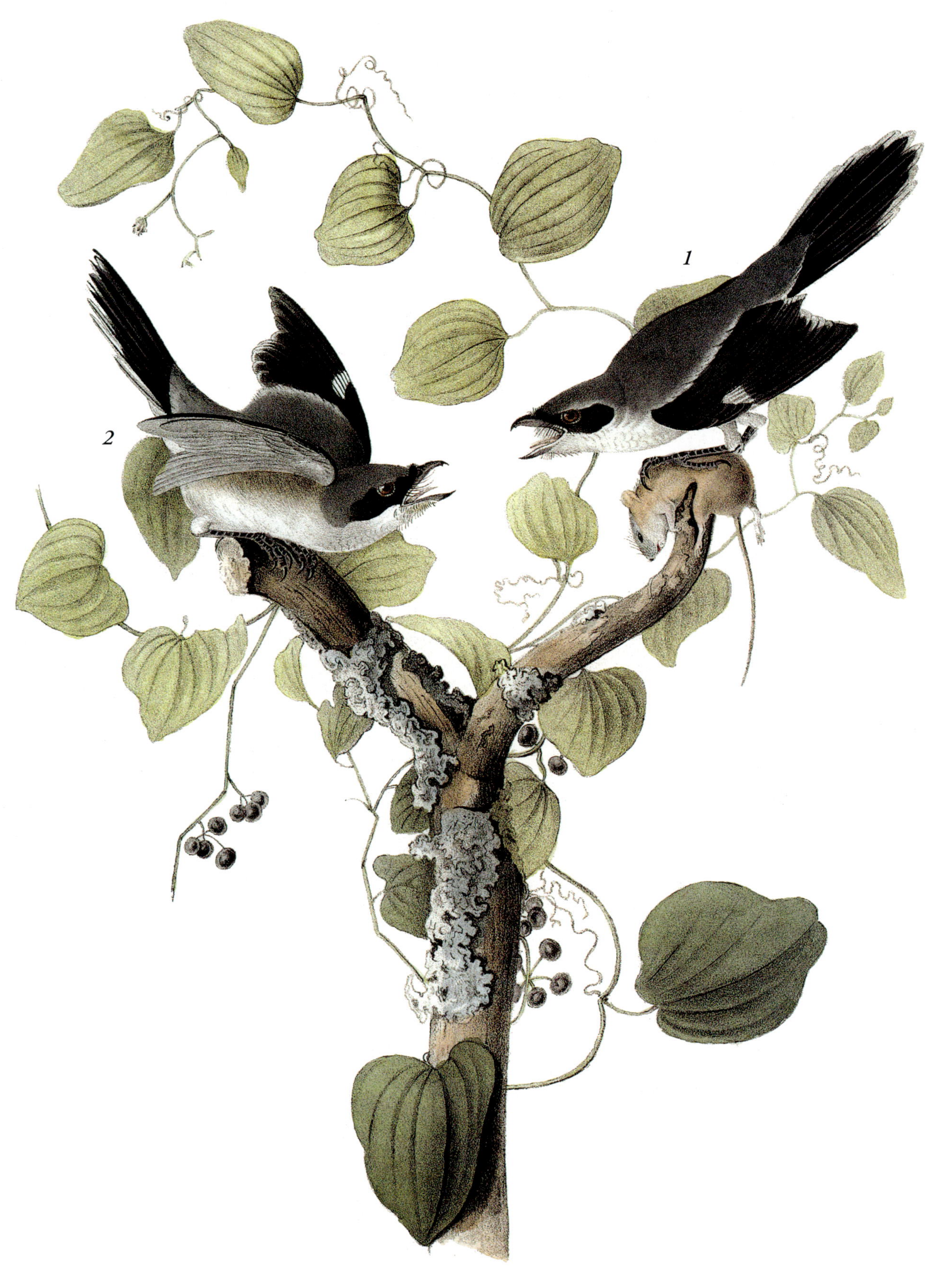

30. Loggerhead Shrike

1. Male 2. Female

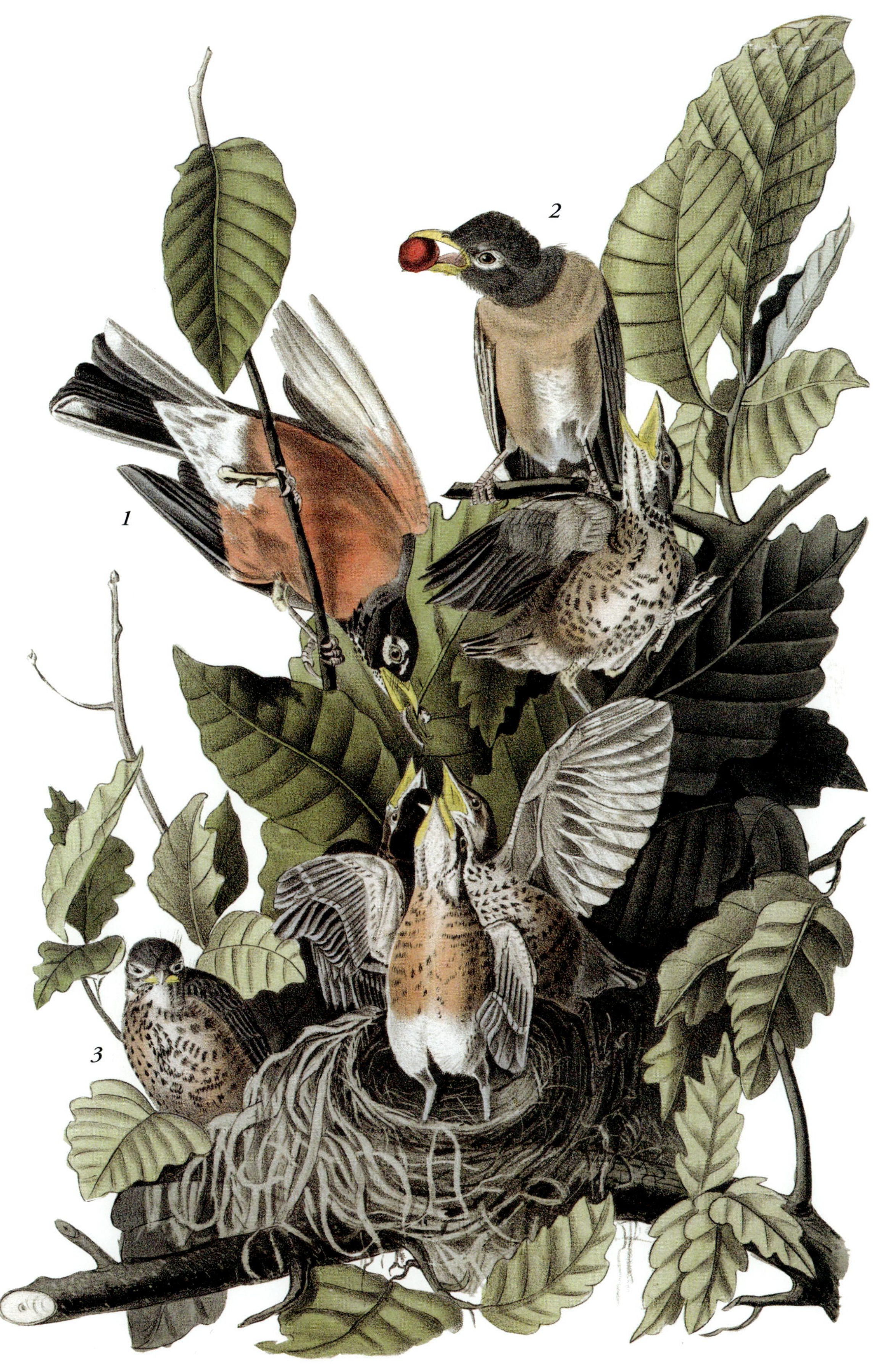

31. American Robin

1. Male 2. Female 3. Young

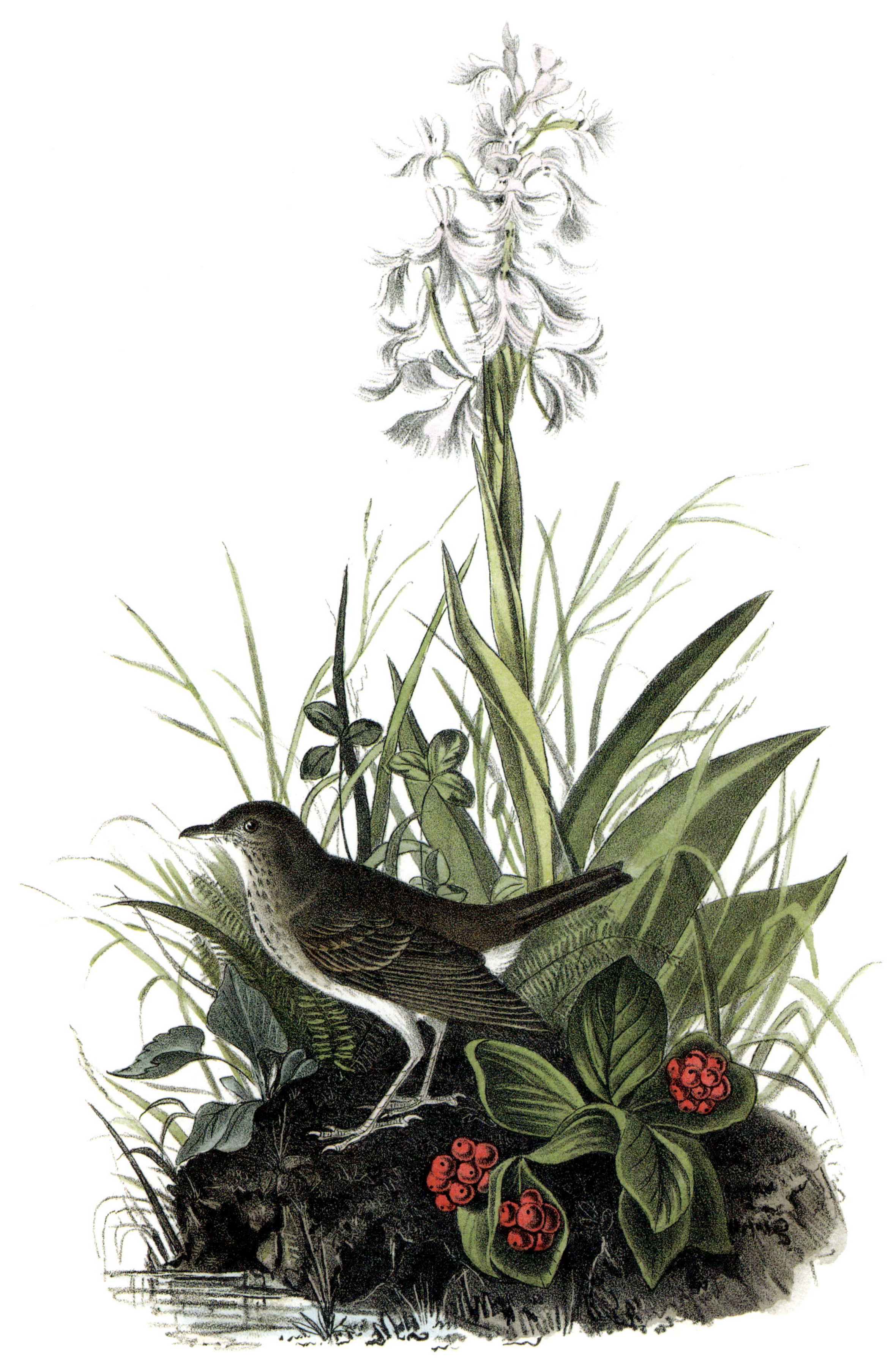

32. Veery

Male

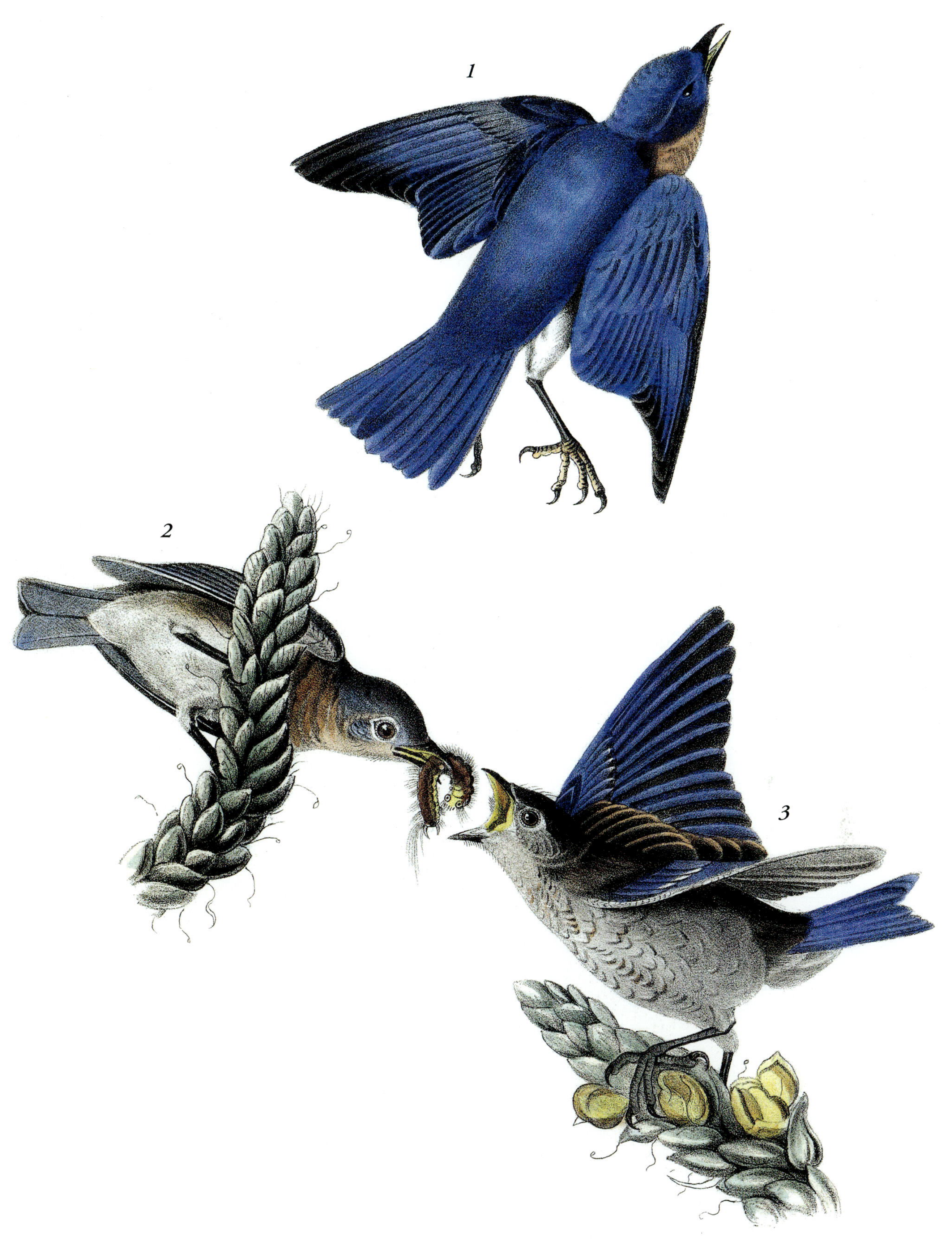

33. Eastern Bluebird

1. Male 2. Female 3. Young

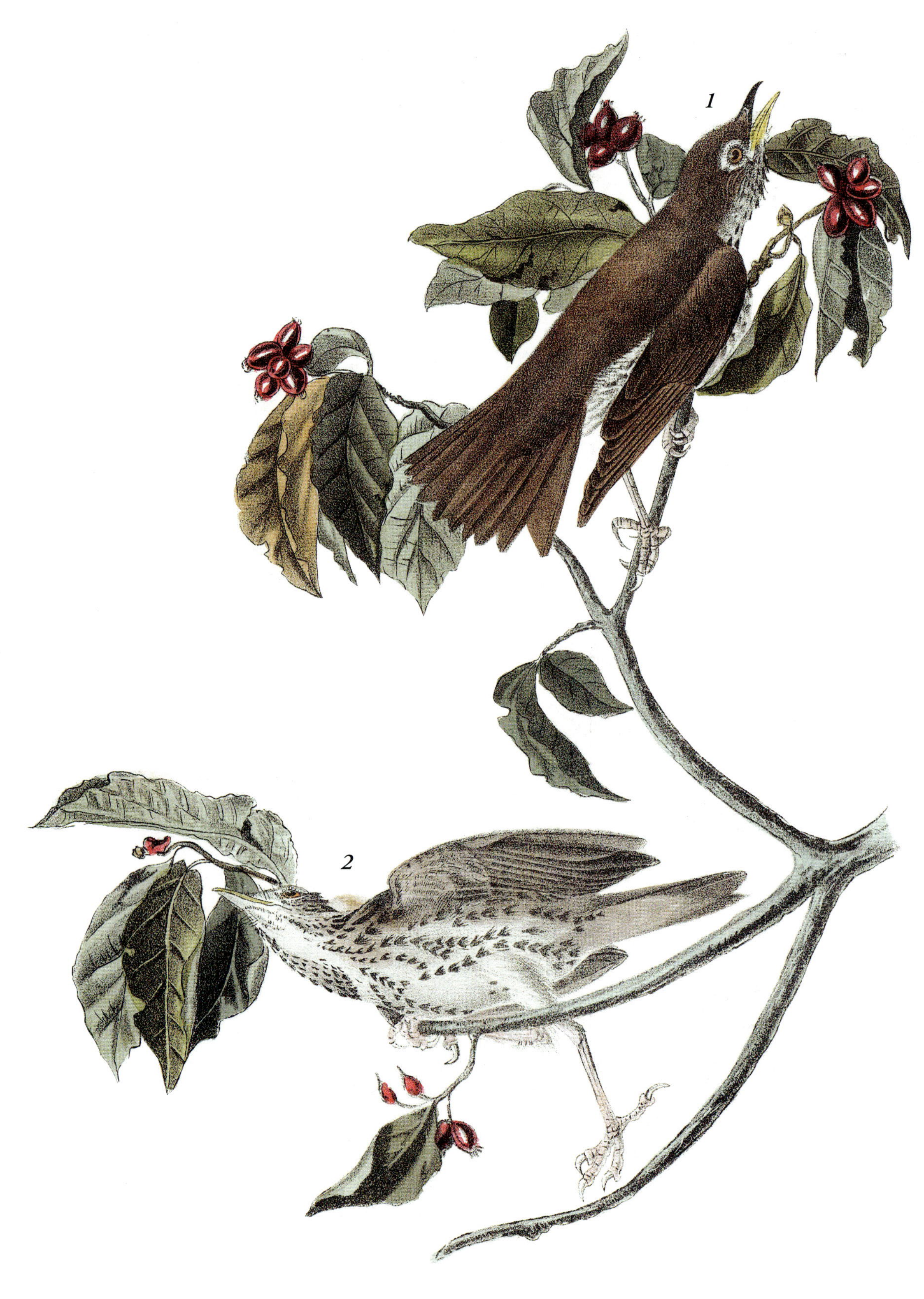

34. Wood Thrush

1. Male 2. Female

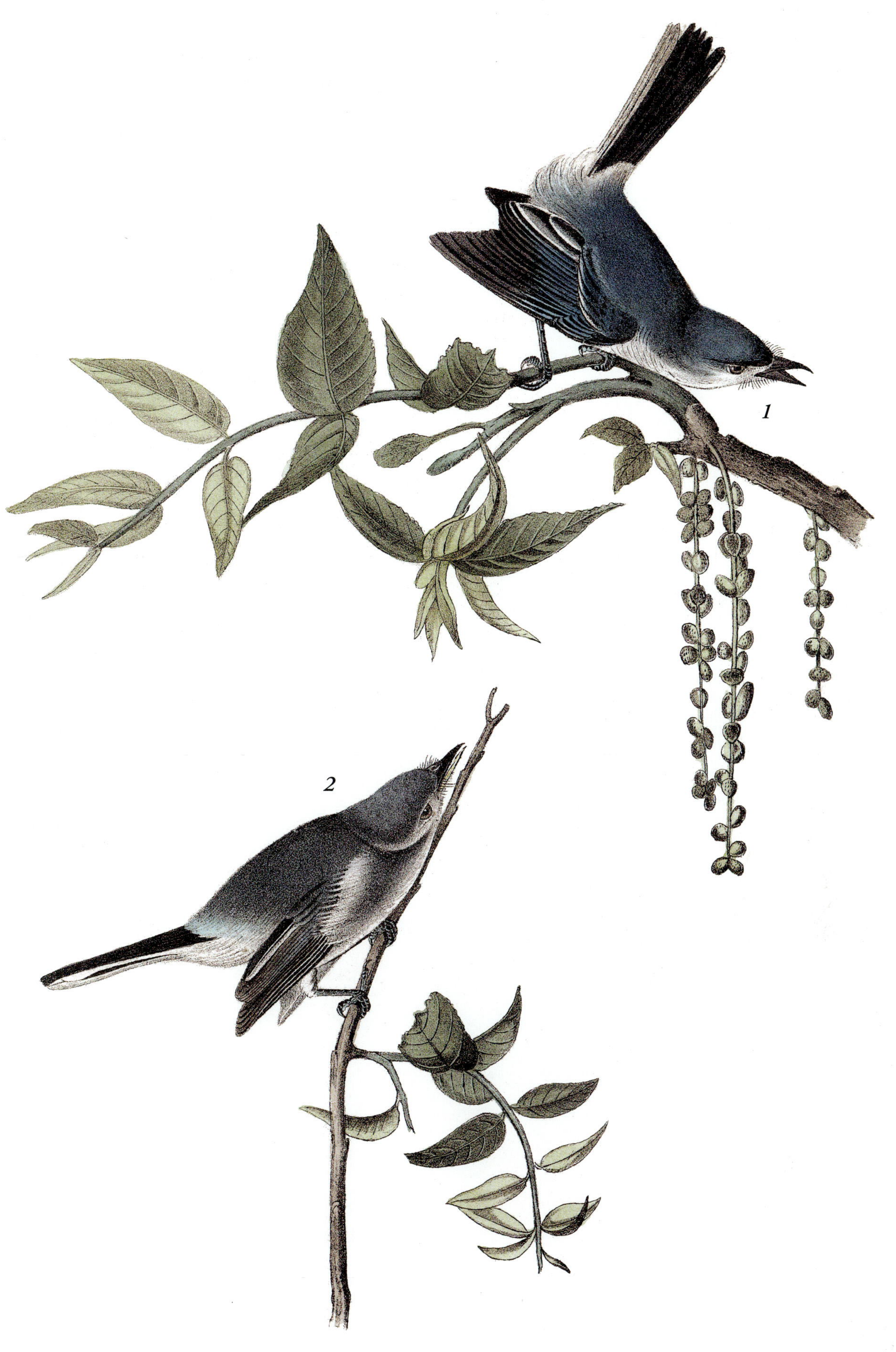

35. Blue-Gray Gnatcatcher

1. Male 2. Female

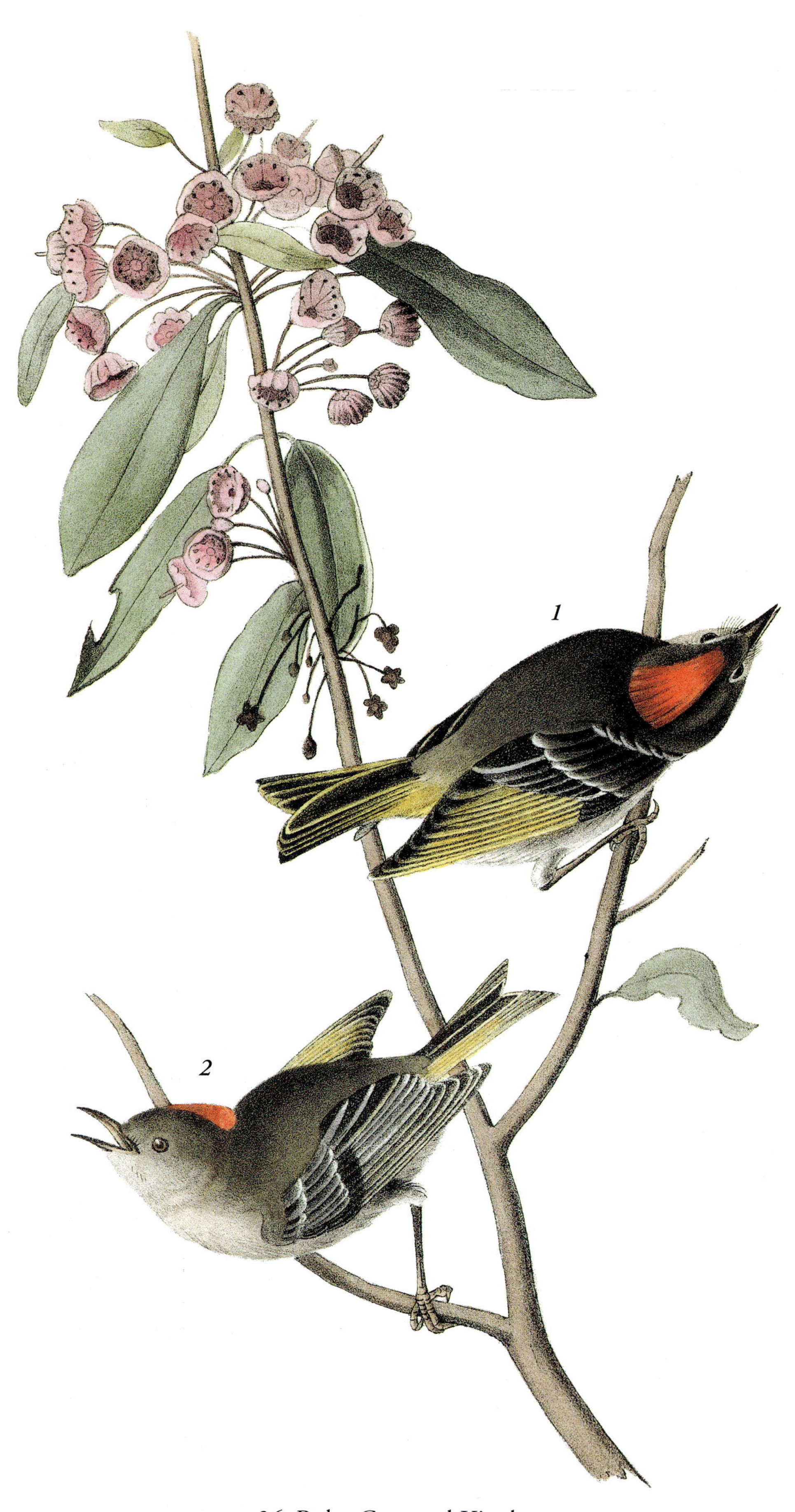

36. Ruby-Crowned Kinglet

1. Male   2. Female

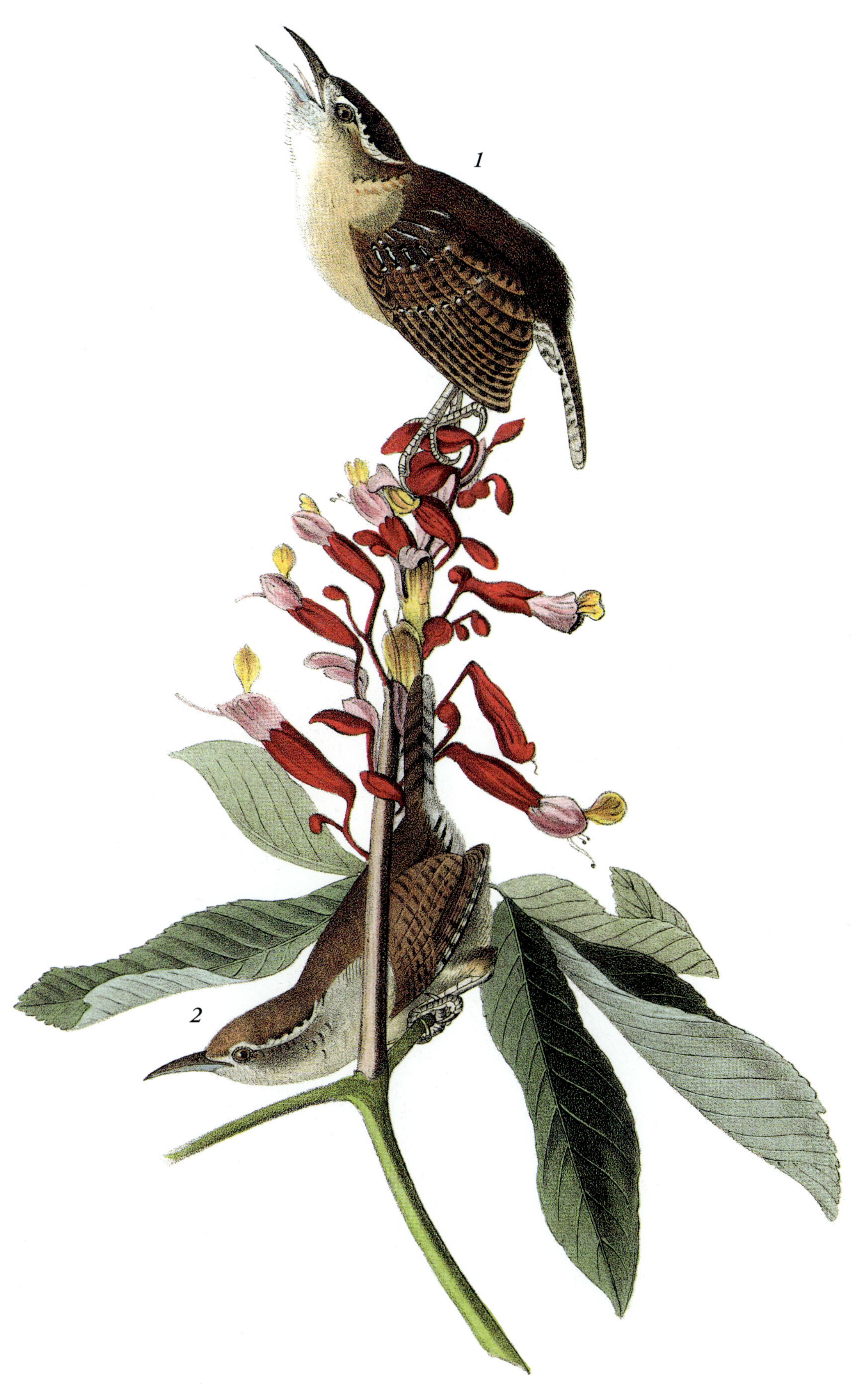

37. Carolina Wren

1. Male 2. Female

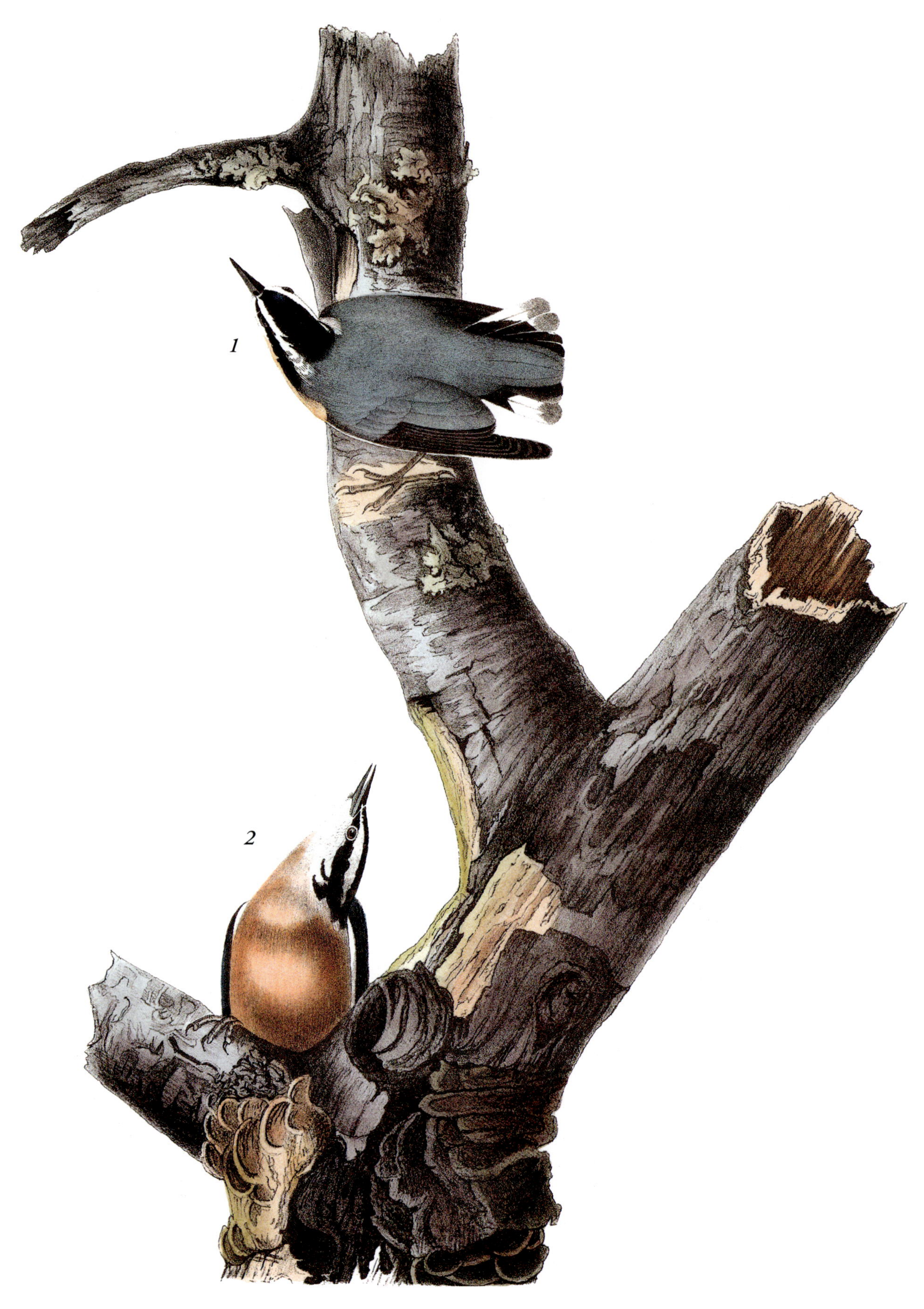

38. Red-Breasted Nuthatch

1. Male 2. Female

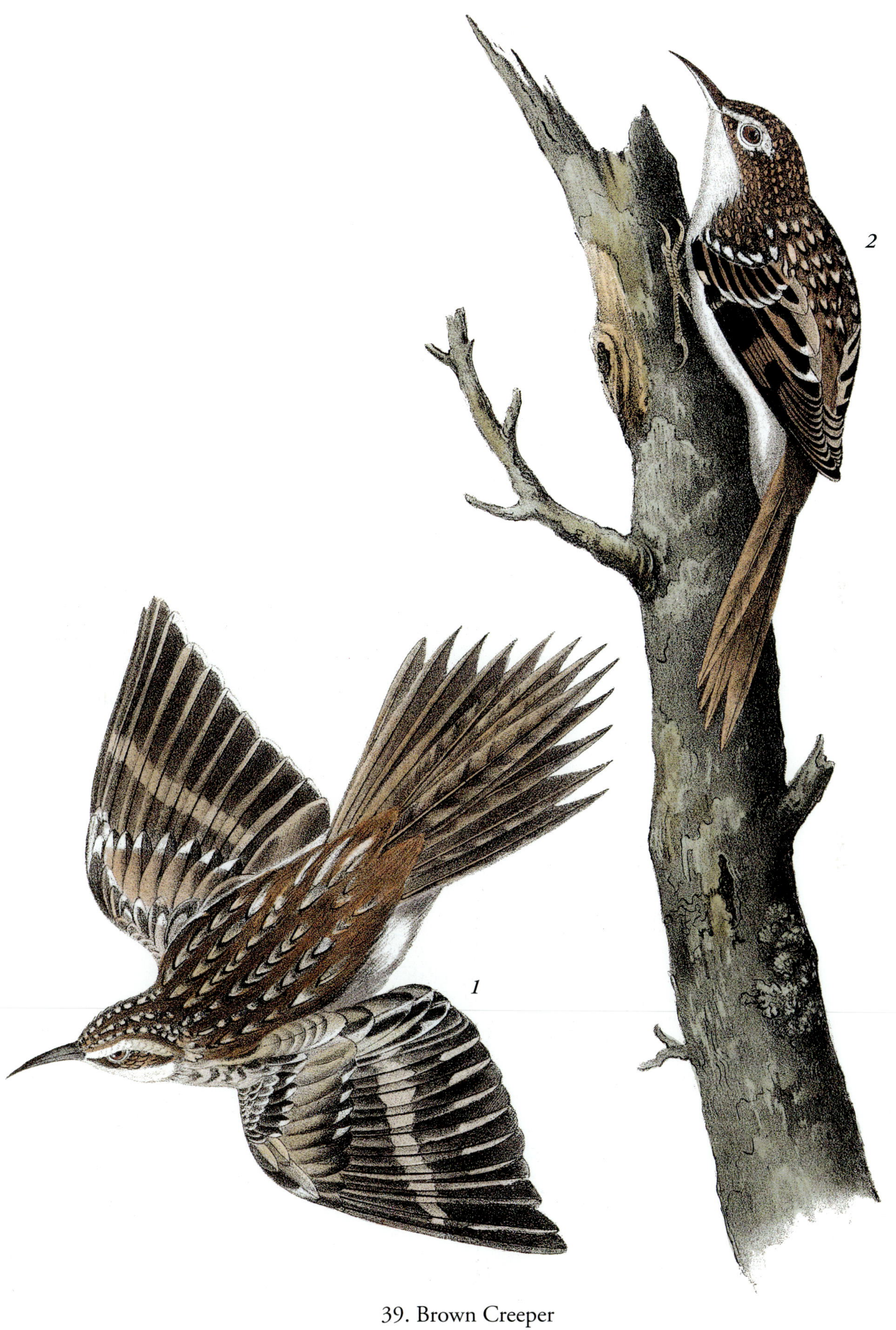

39. Brown Creeper

1. Male 2. Female

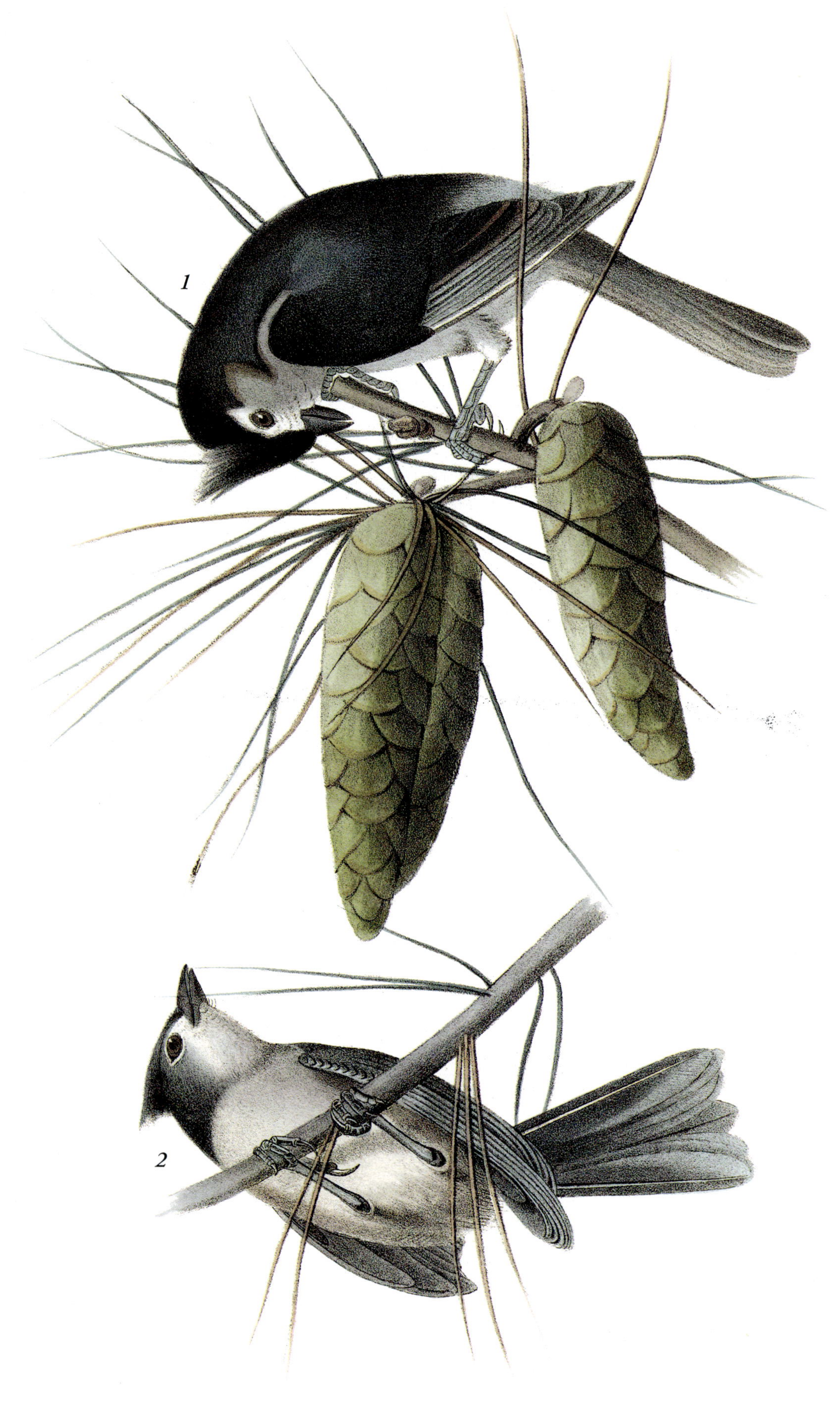

40. Tufted Titmouse

1. Male 2. Female

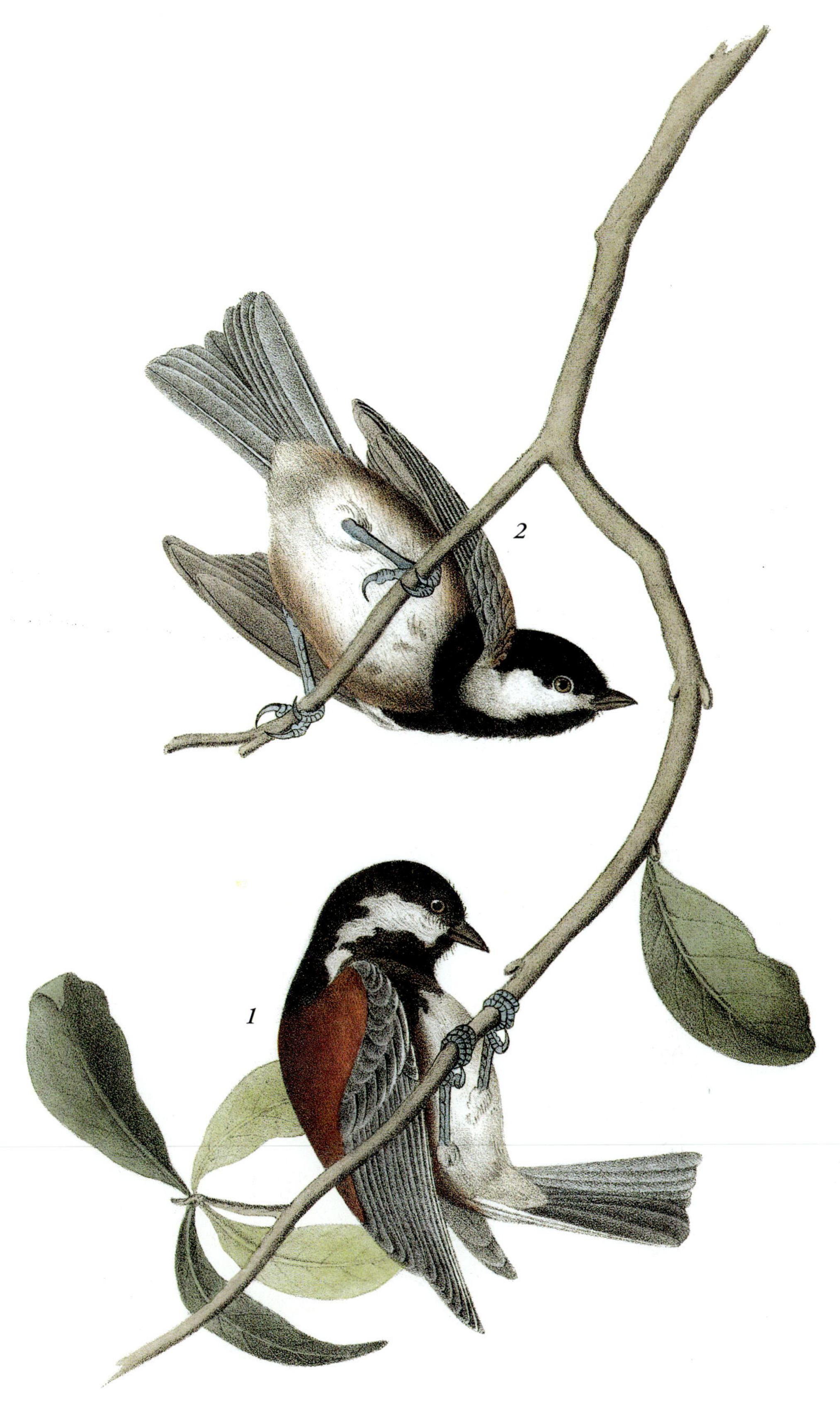

41. Chestnut-Backed Chickadee

1. Male 2. Female

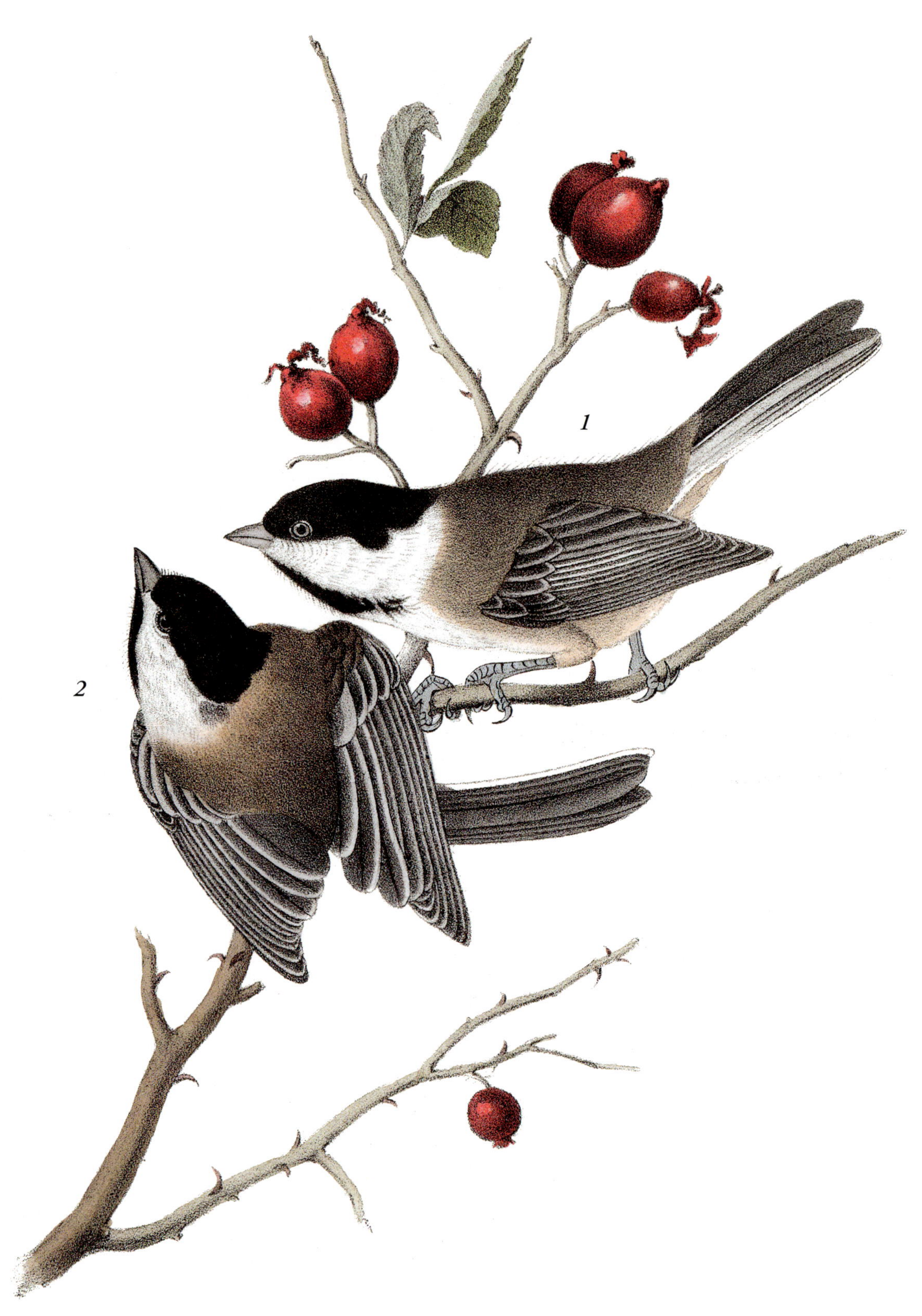

42. Black-Capped Chickadee

1. Male 2. Female

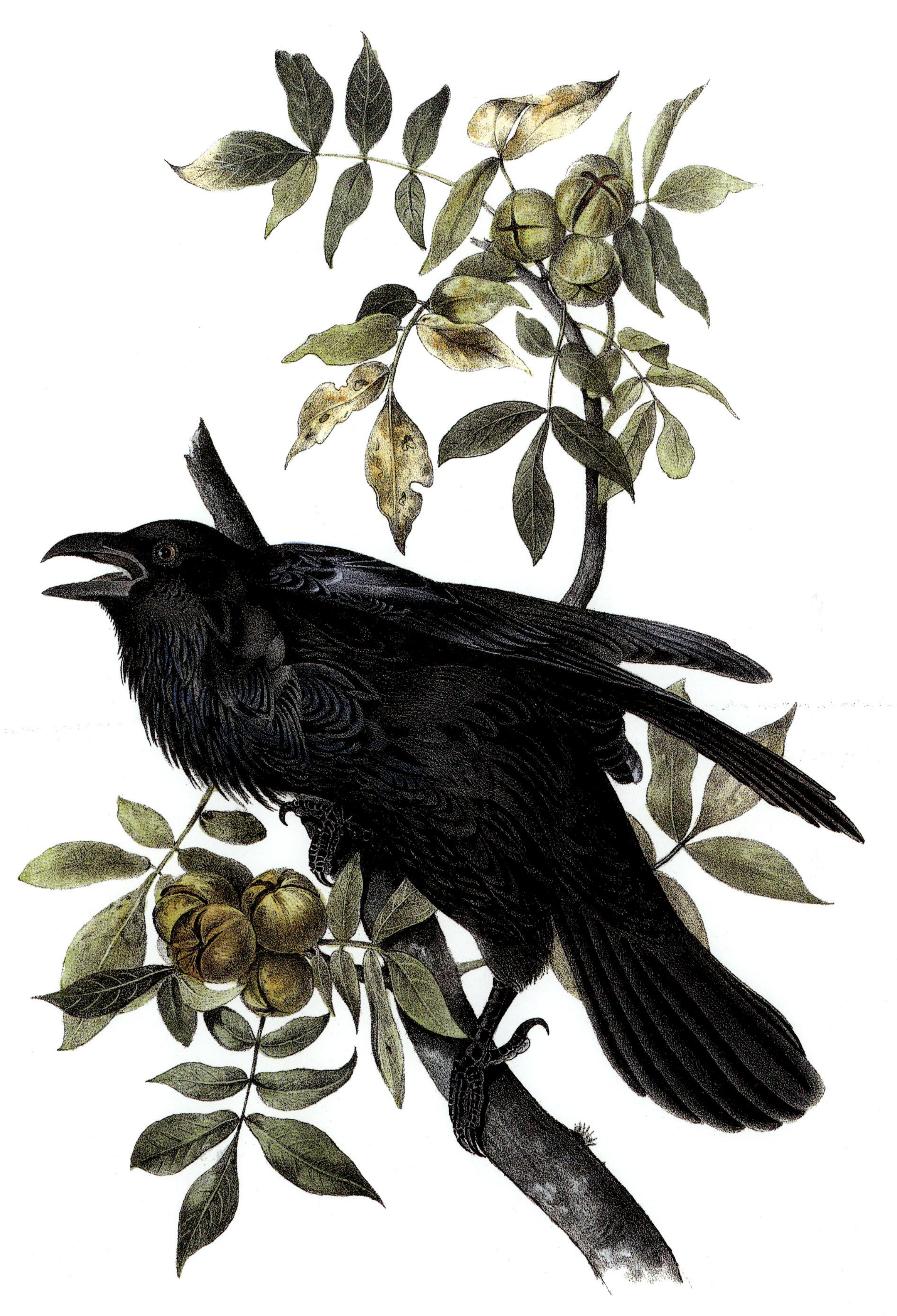

43. Common Raven

Old Male

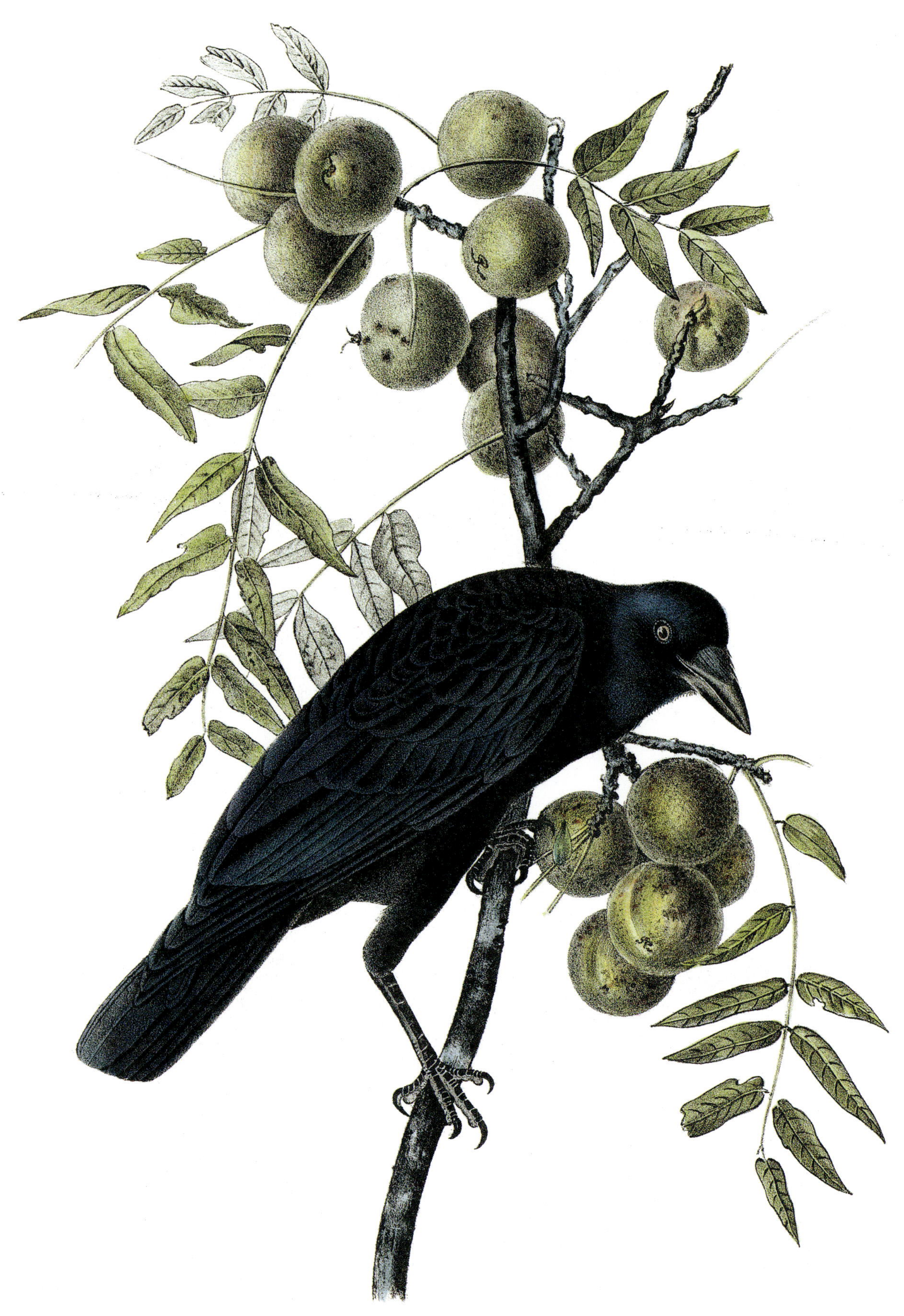

44. American Crow
Male

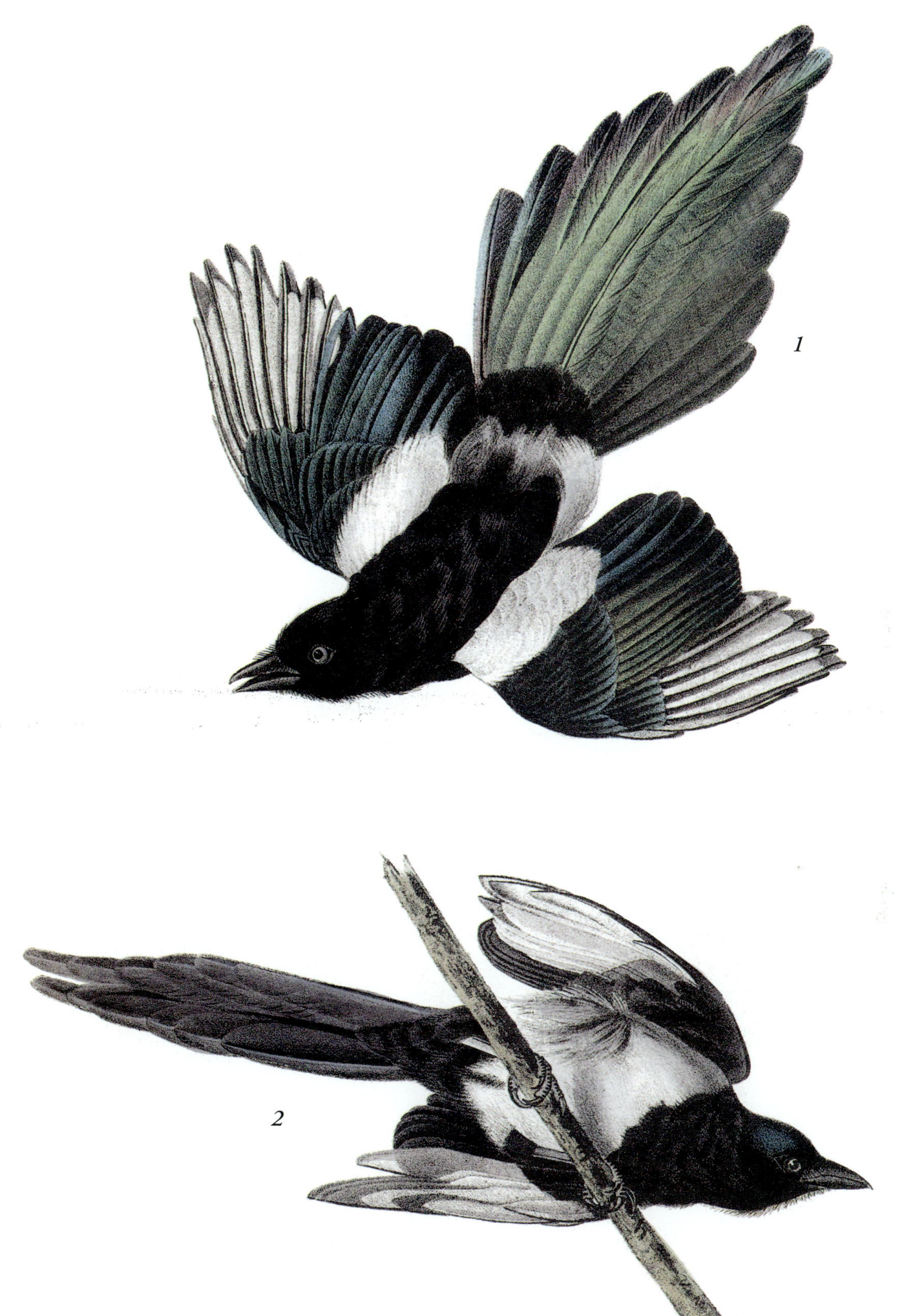

45. Black-Billed Magpie

1. Male 2. Female

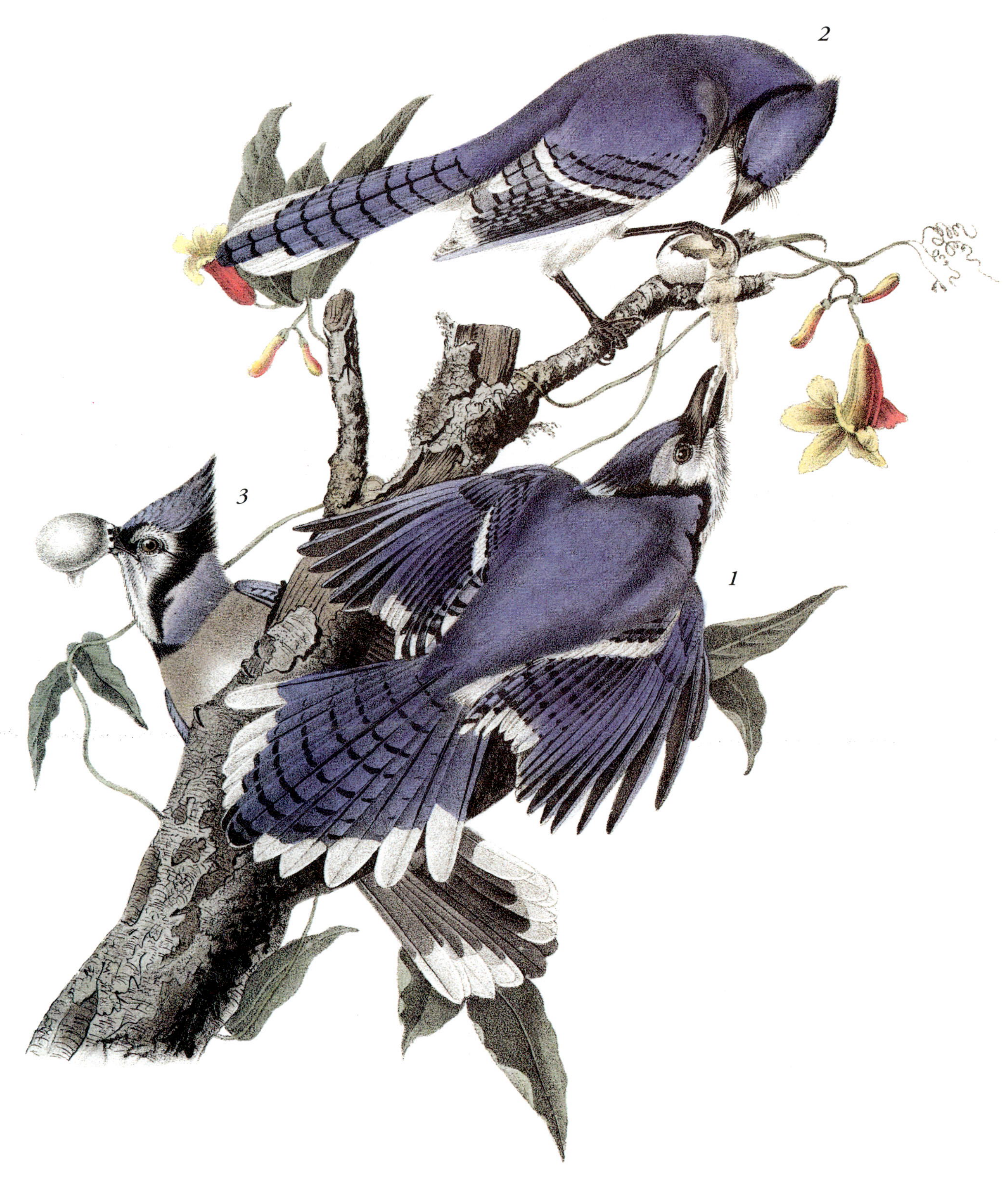

46. Blue Jay

1. Male 2, 3. Females

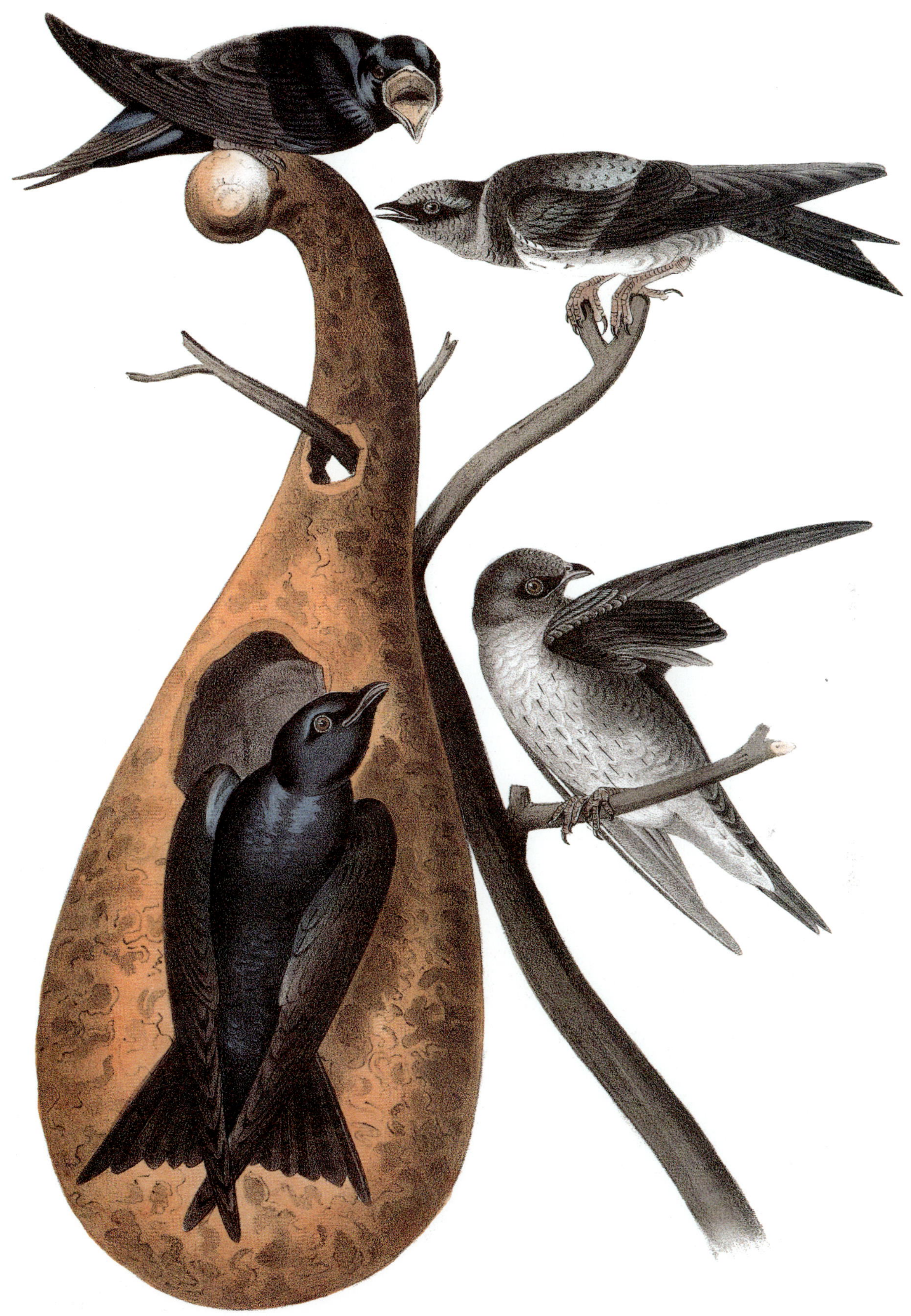

47. Purple Martin

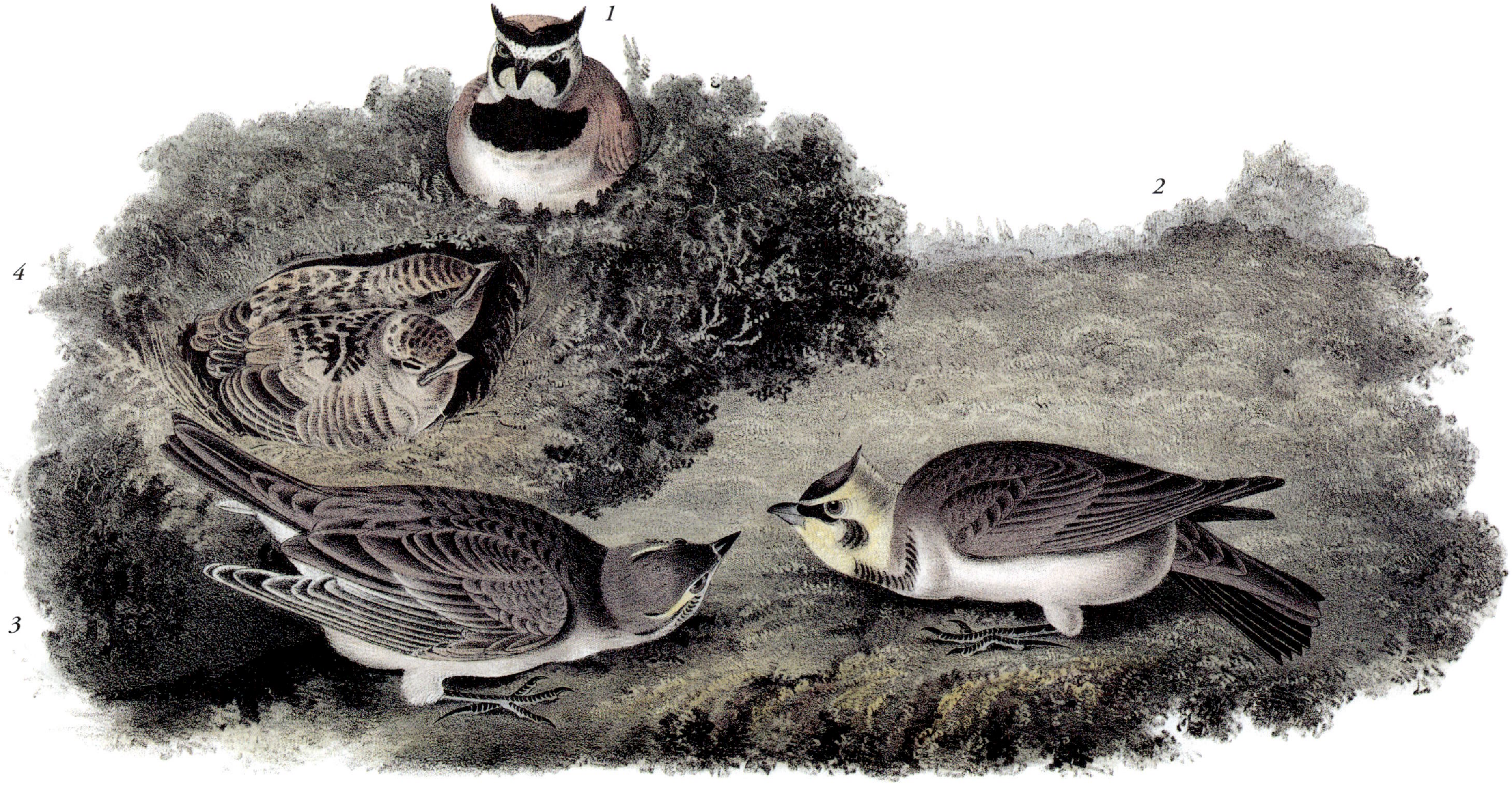

48. Horned Lark

1. Male, Summer Plumage 2. Male, Winter Plumage 3. Female 4. Young and Nest

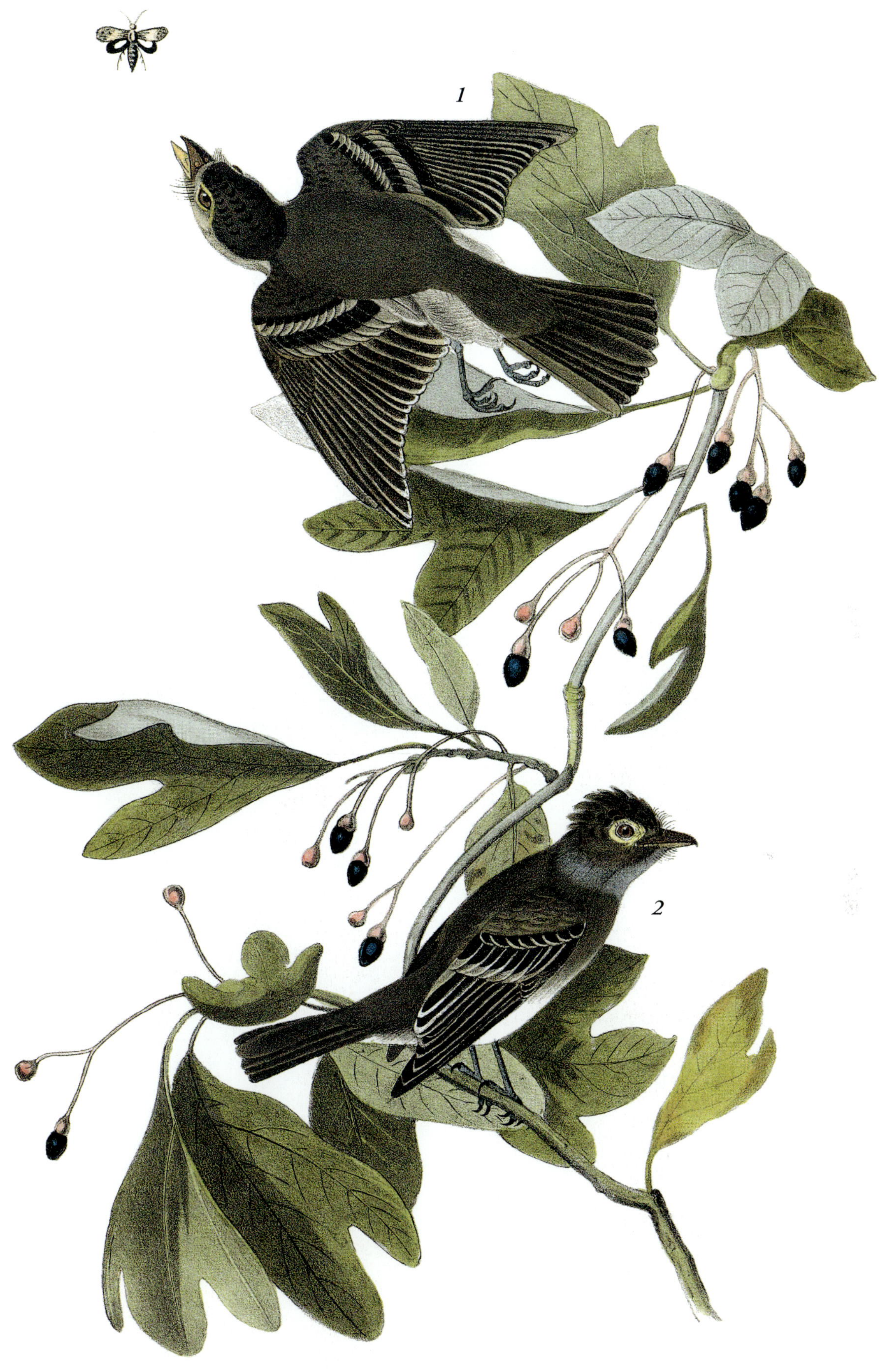

49. Acadian Flycatcher

1. Male 2. Female

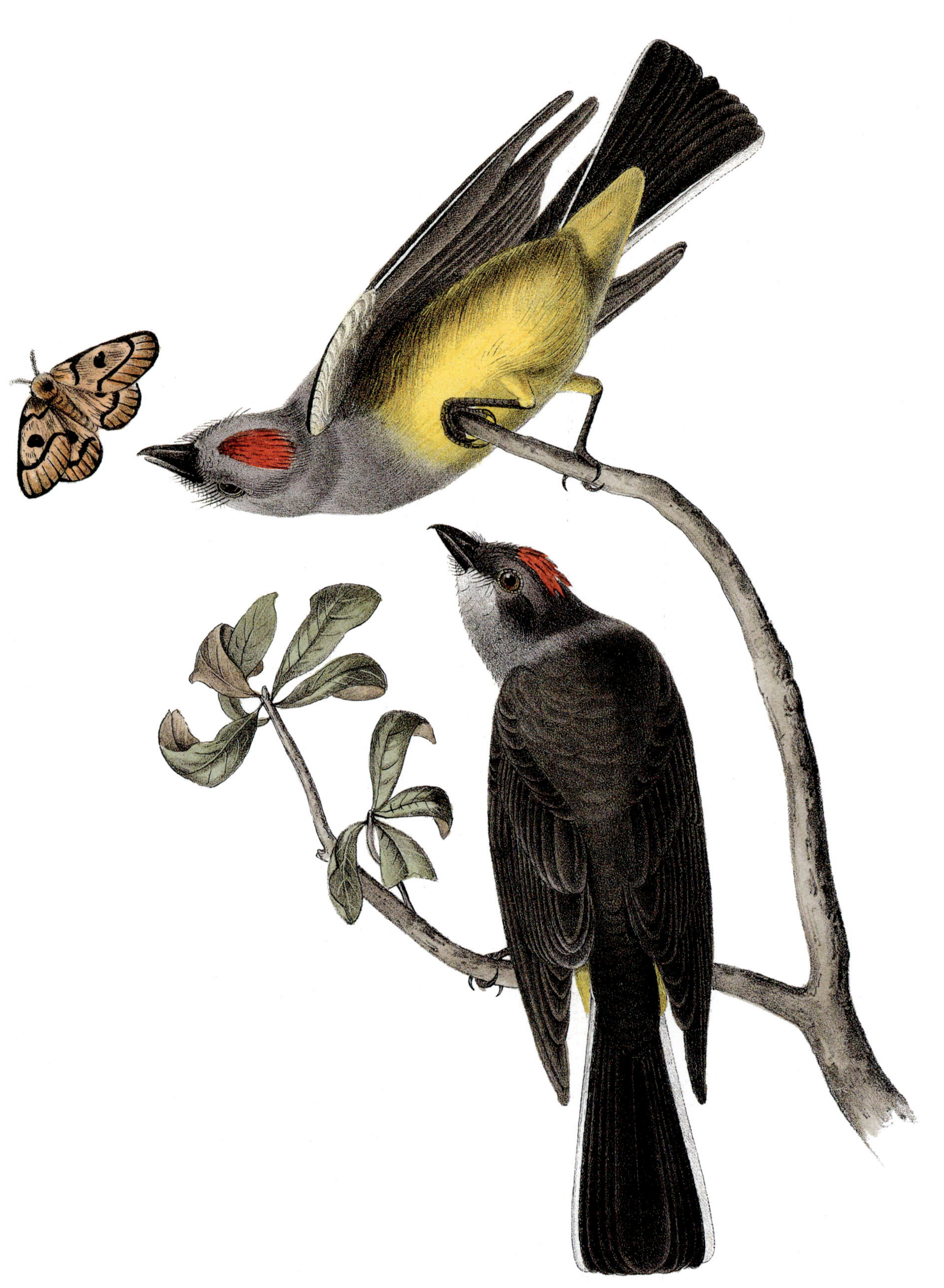

50. Western Kingbird

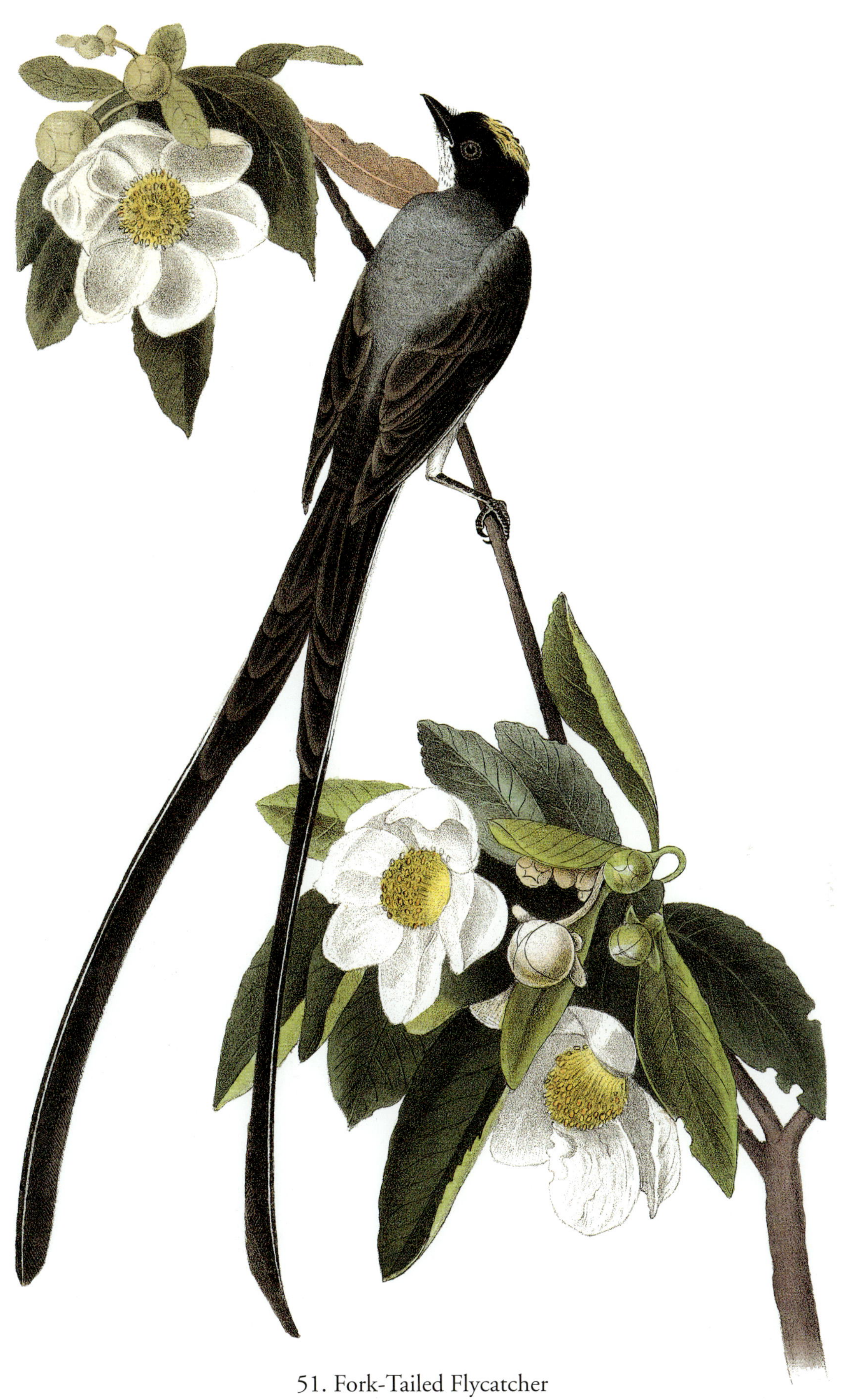

51. Fork-Tailed Flycatcher

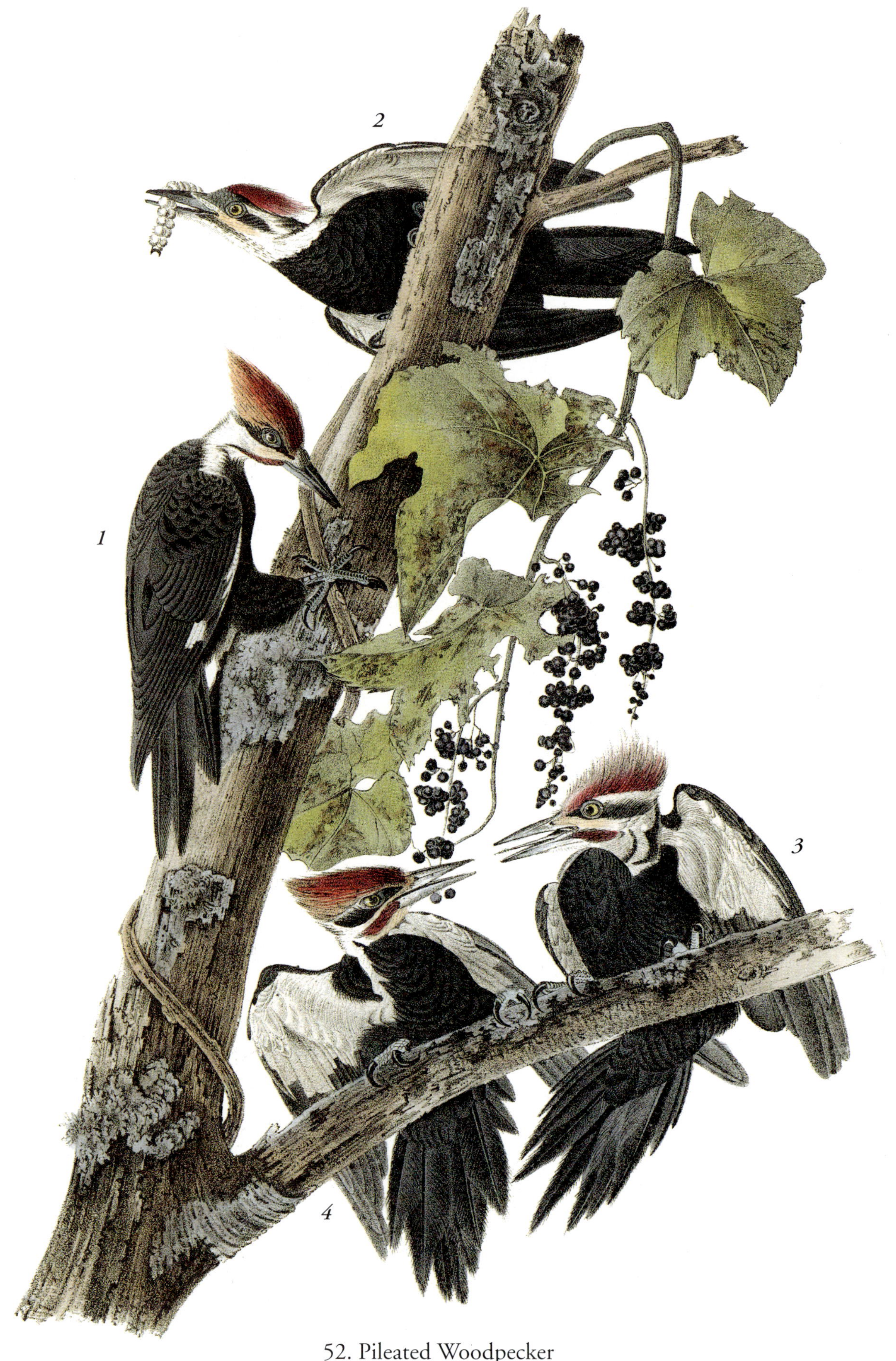

52. Pileated Woodpecker

1. Adult Male 2. Adult Female 3, 4. Young Males

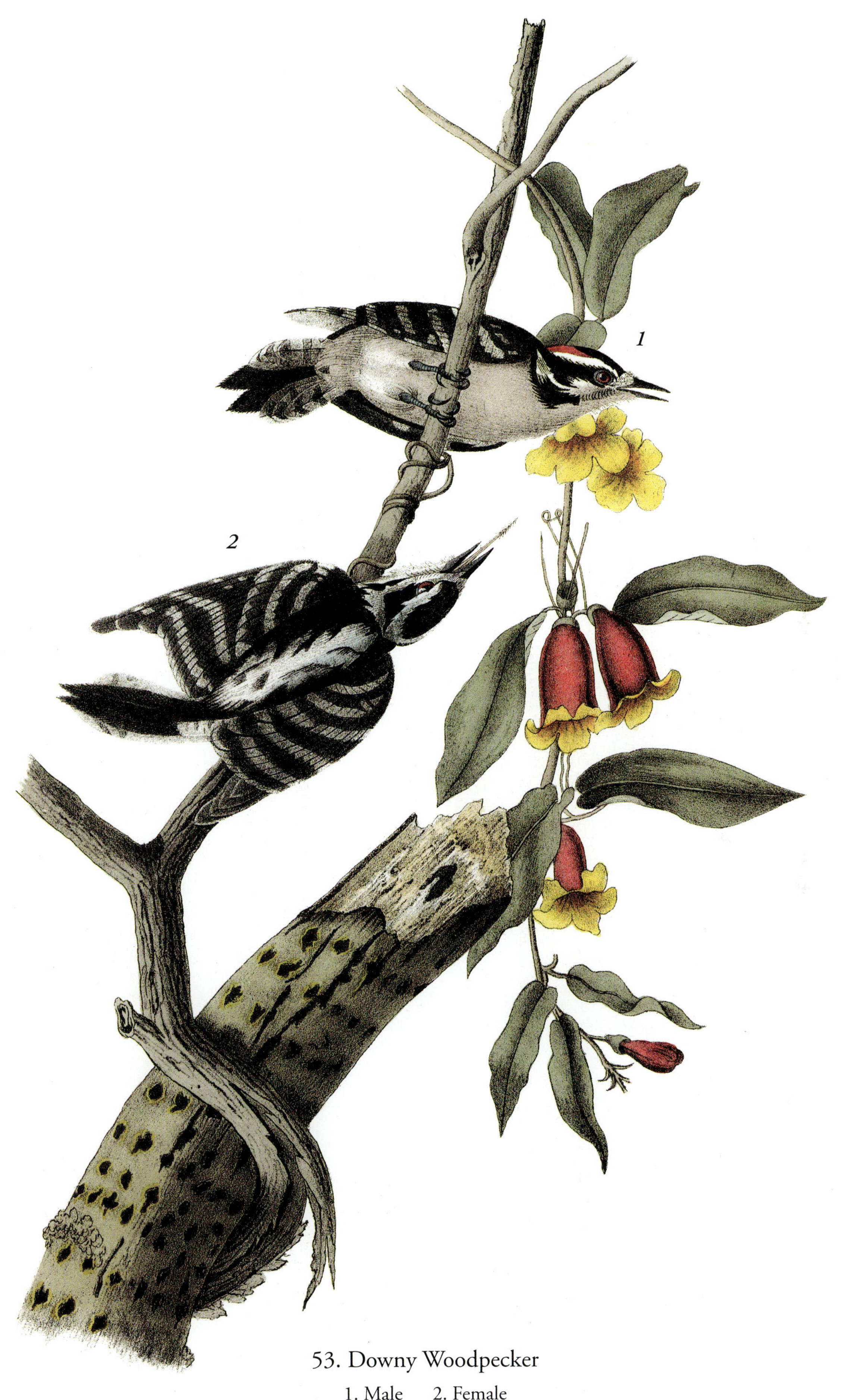

53. Downy Woodpecker

1. Male 2. Female

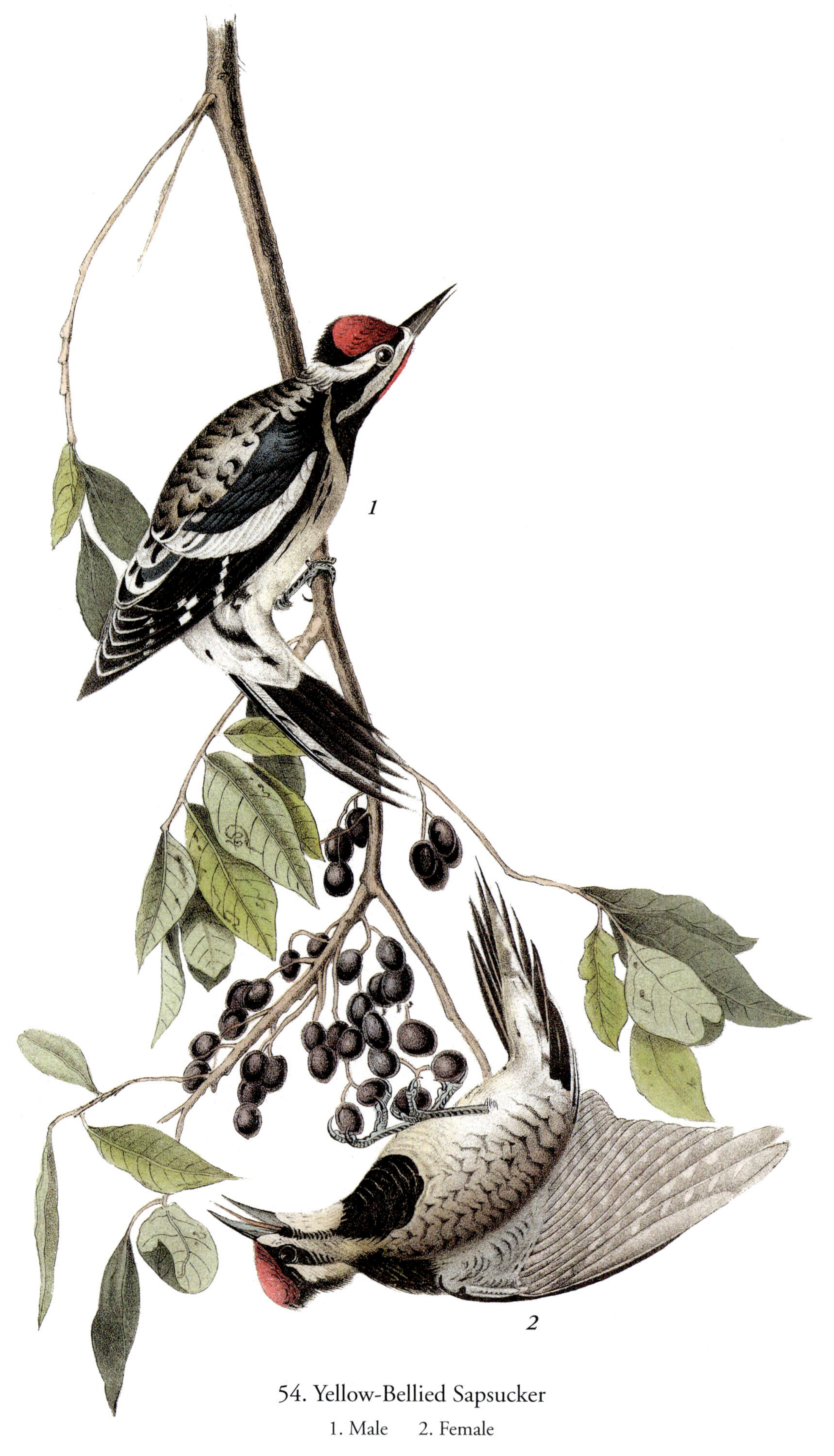

54. Yellow-Bellied Sapsucker

1. Male 2. Female

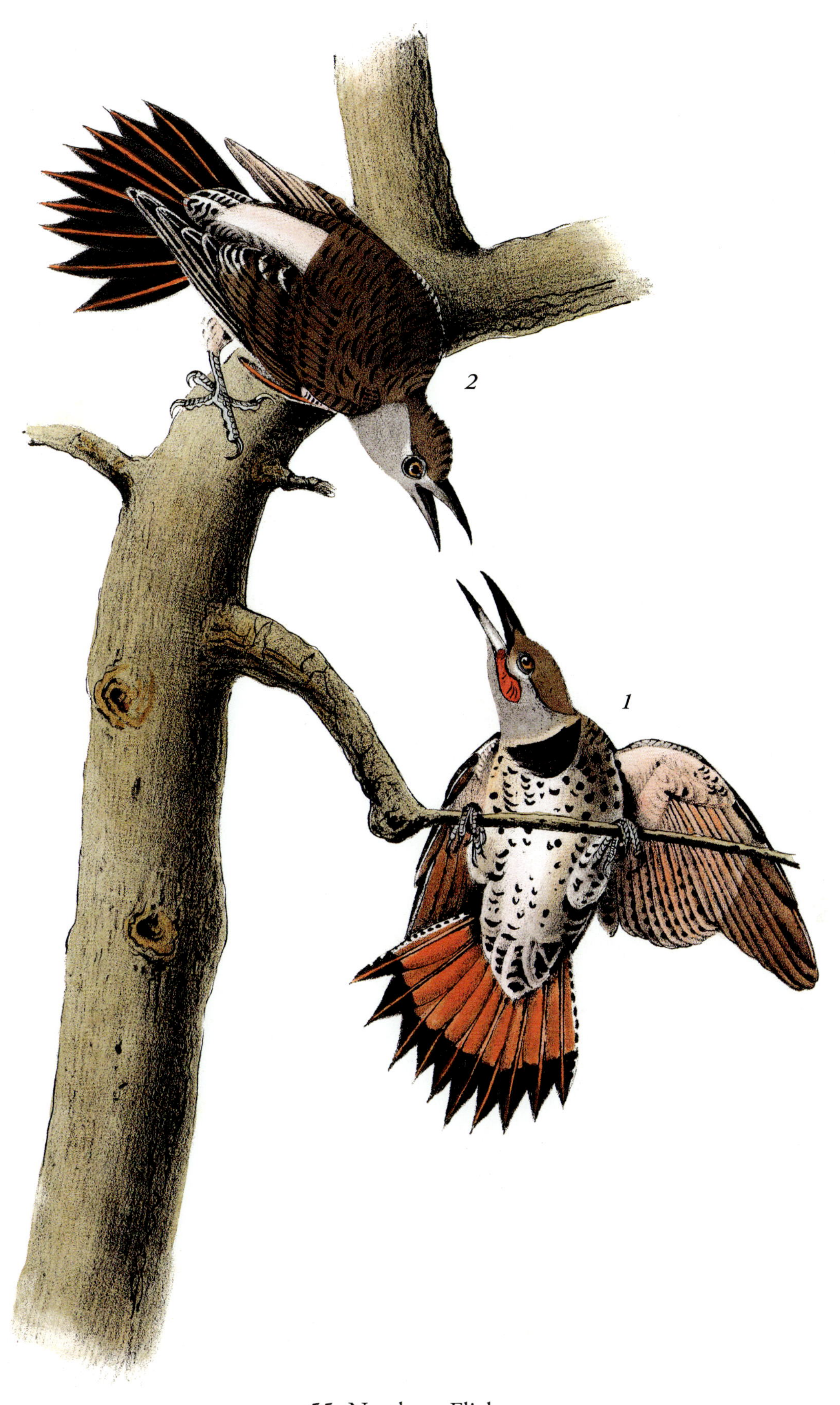

55. Northern Flicker

1. Male 2. Female

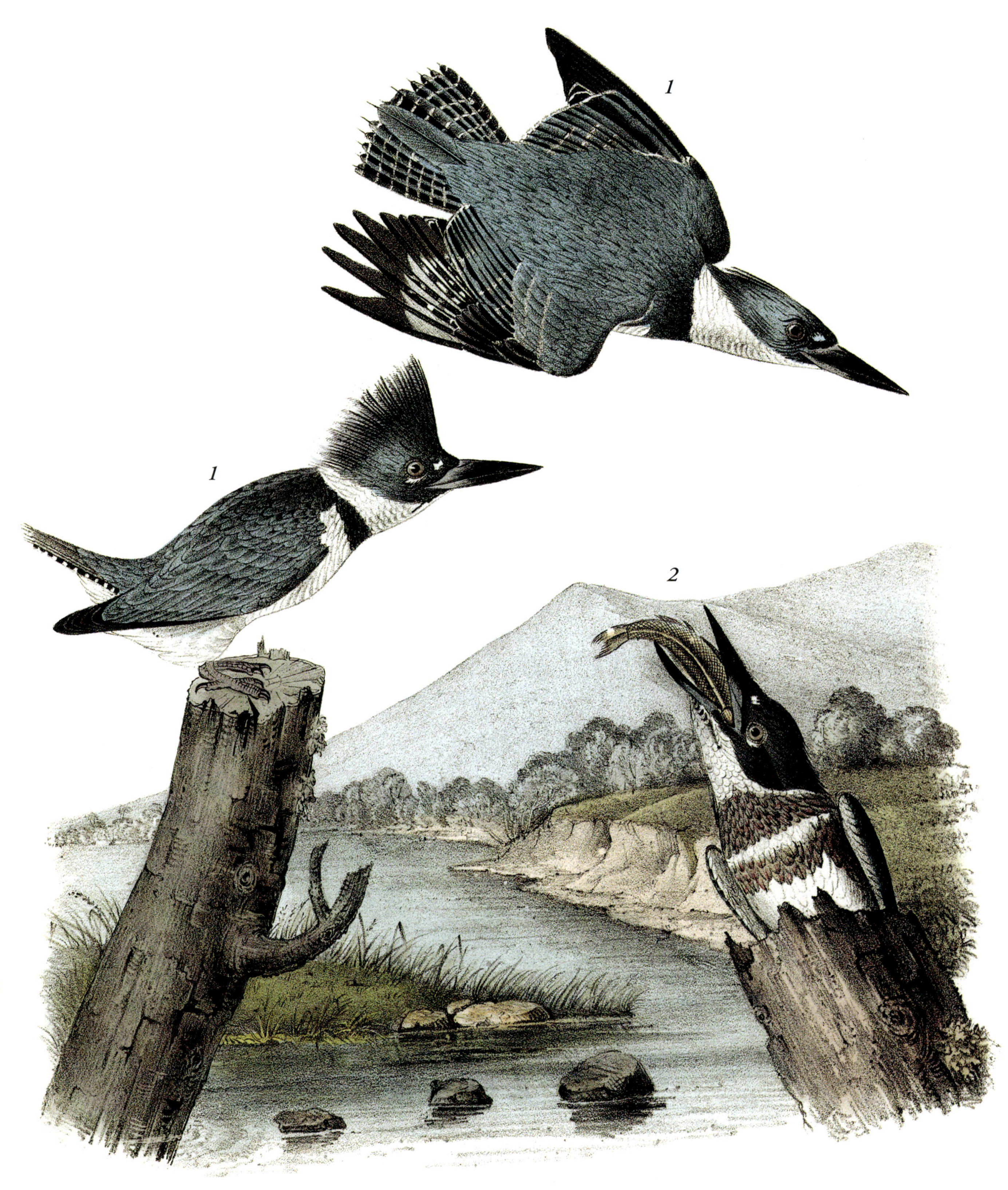

56. Belted Kingfisher

1. Males 2. Female

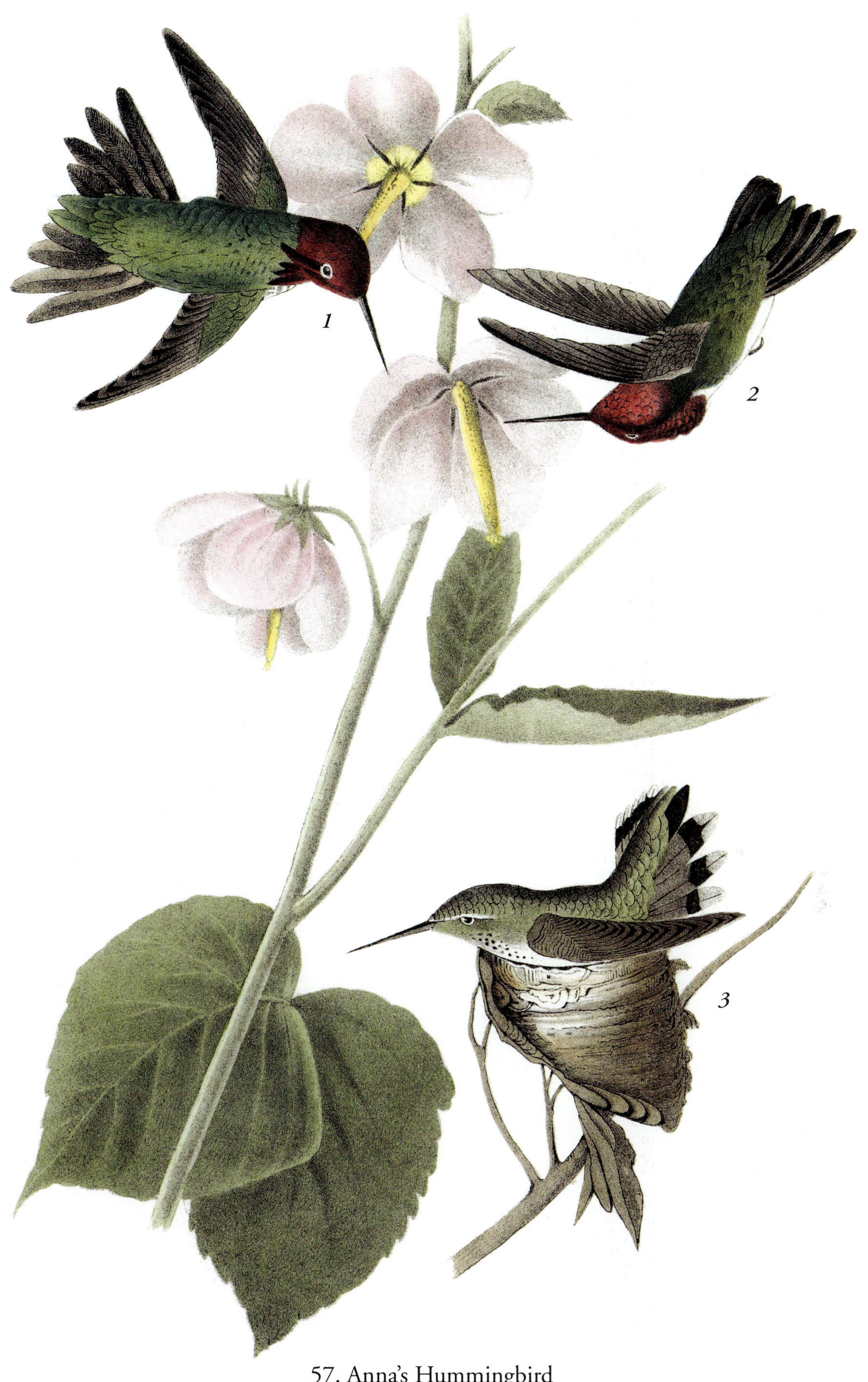

57. Anna's Hummingbird

1, 2. Males 3. Female

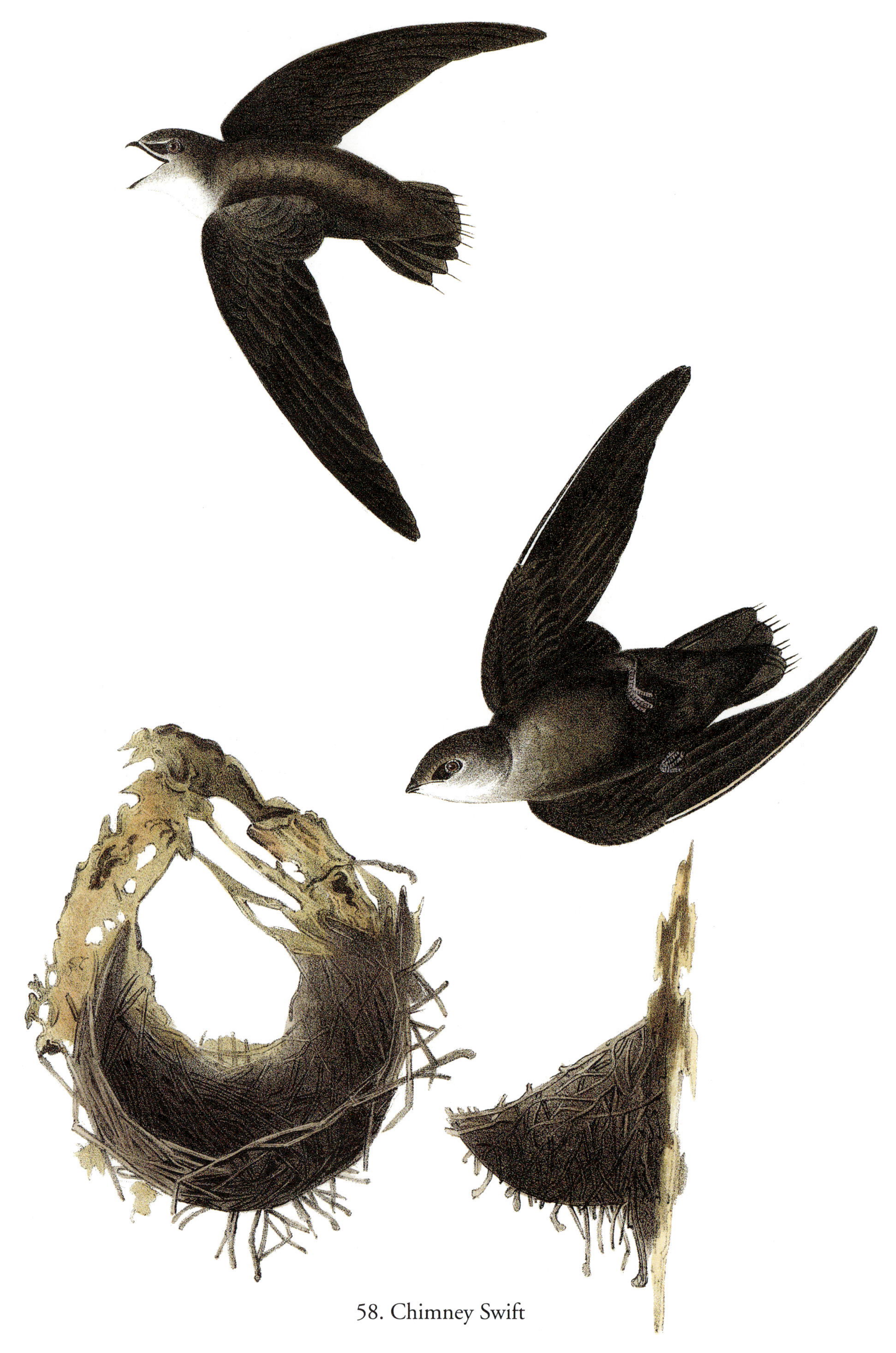

58. Chimney Swift

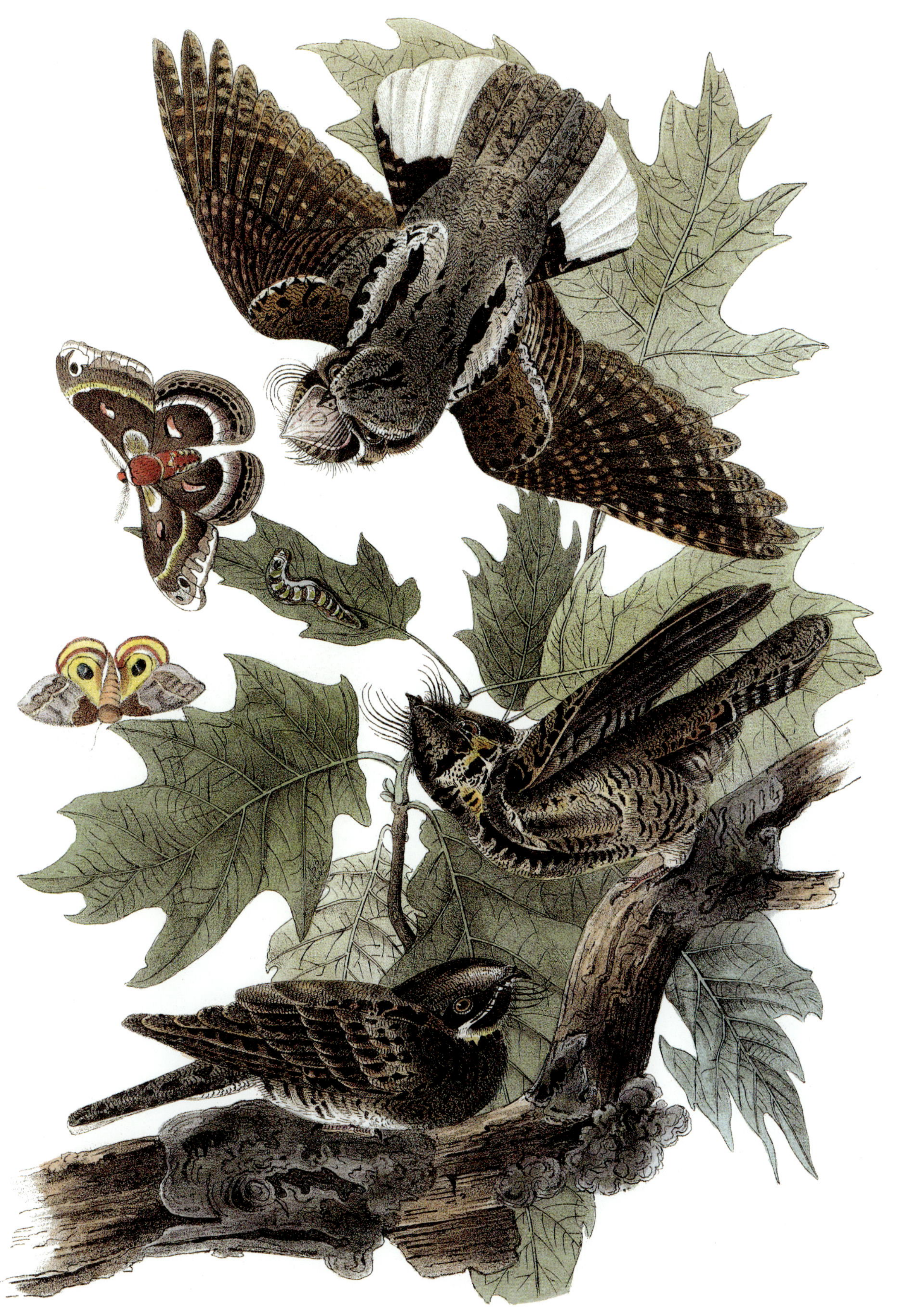

59. Whip-poor-will

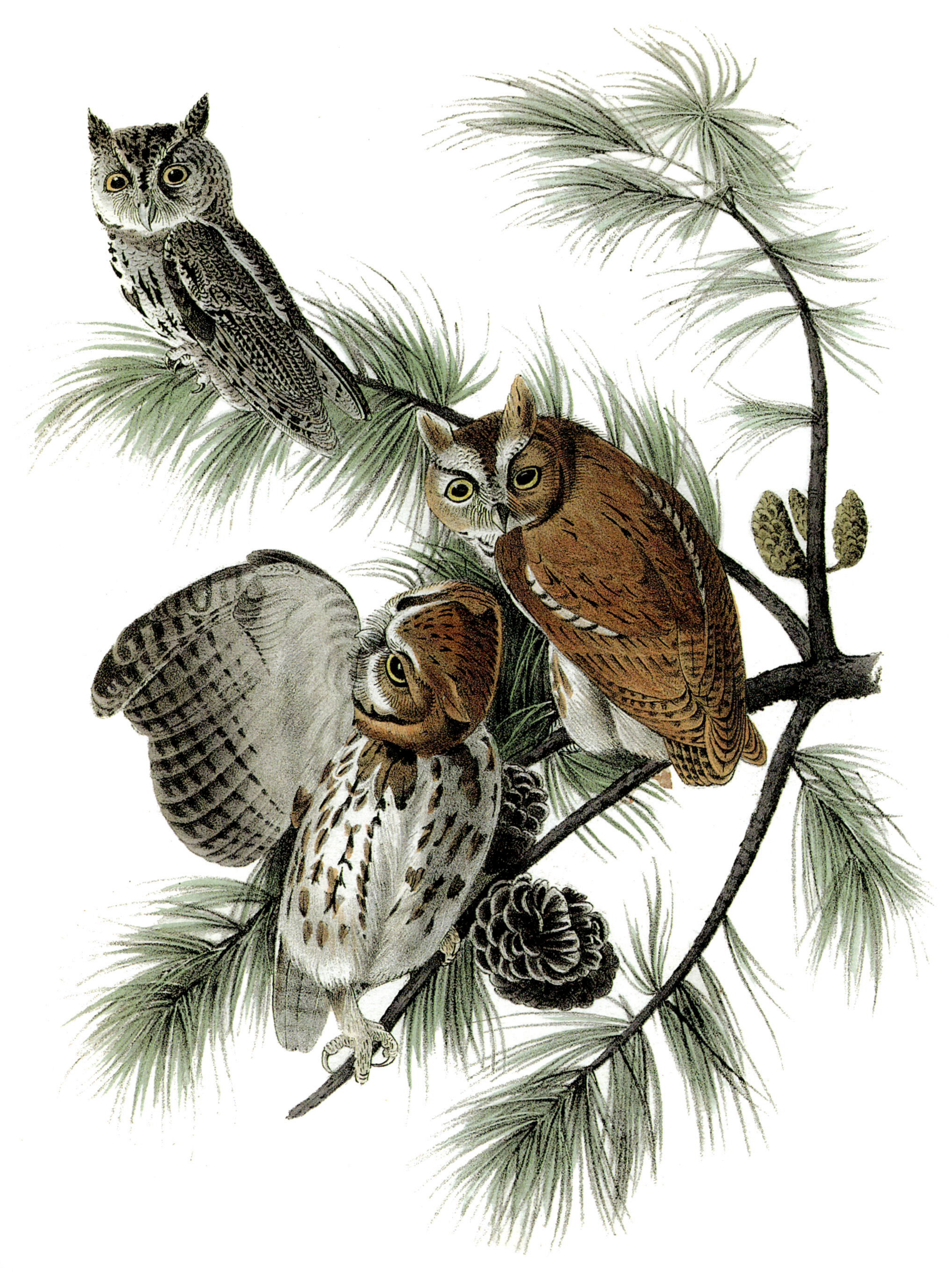

60. Eastern Screech Owl

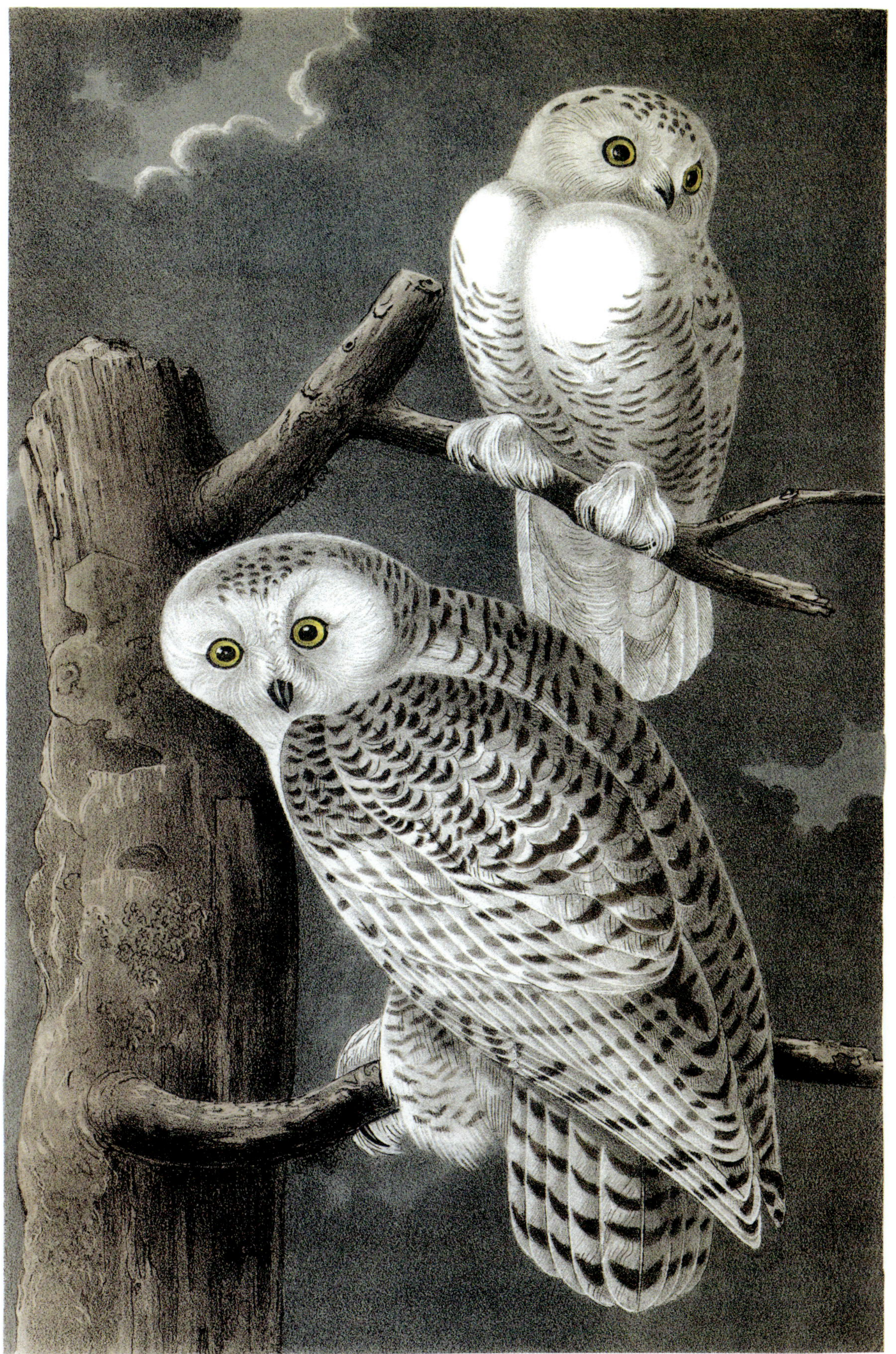

61. Snowy Owl

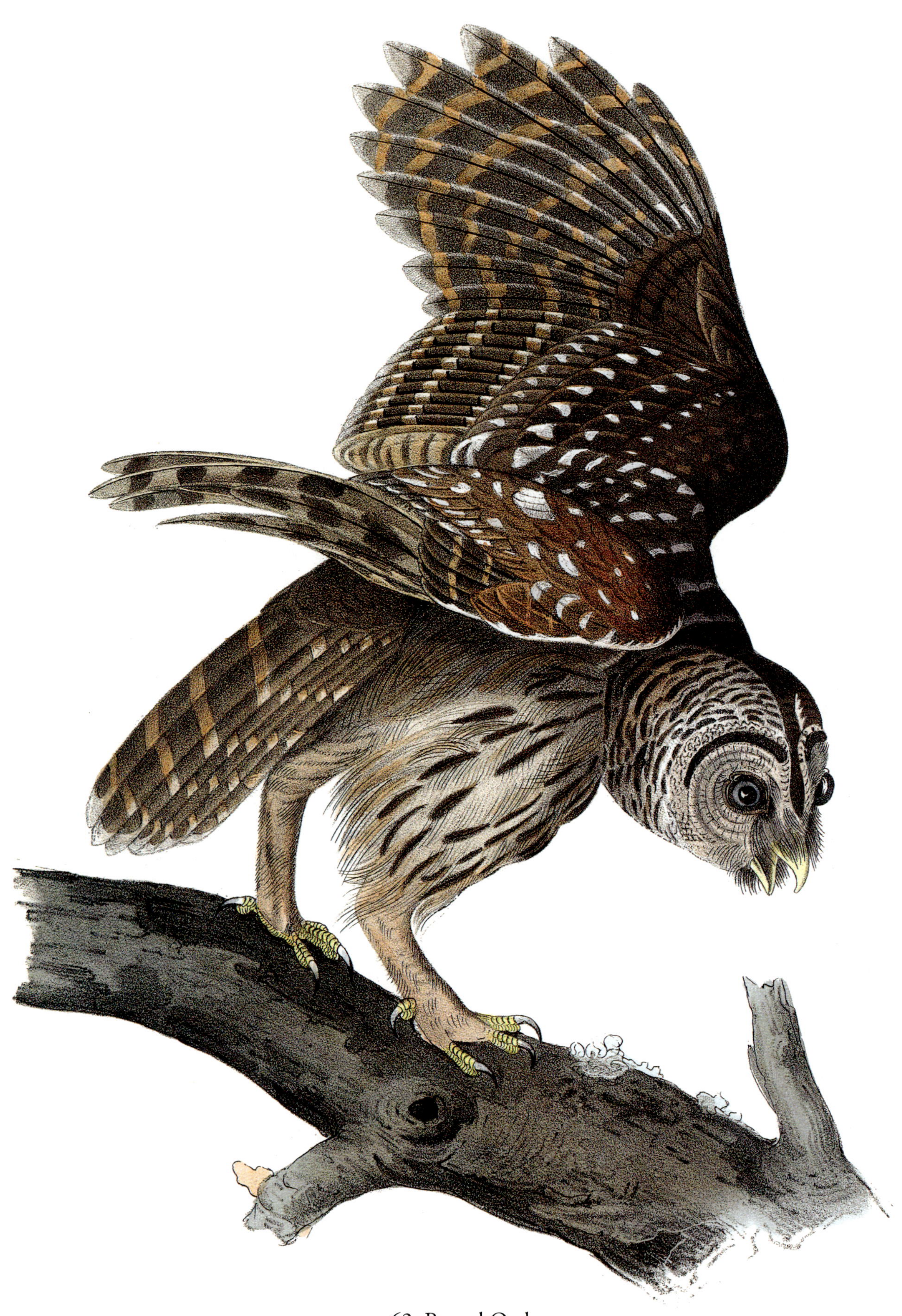

62. Barred Owl

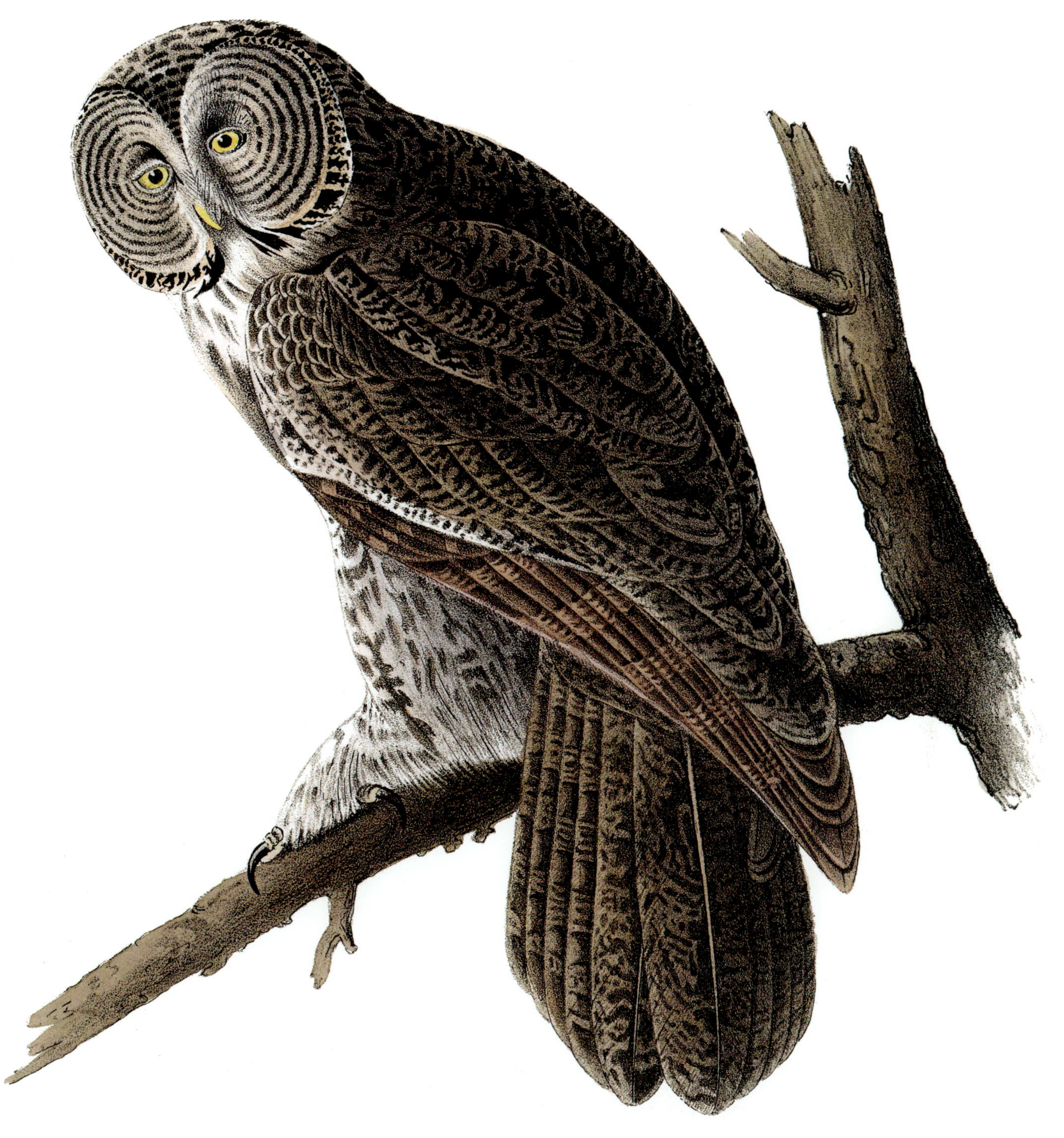

63. Great Gray Owl

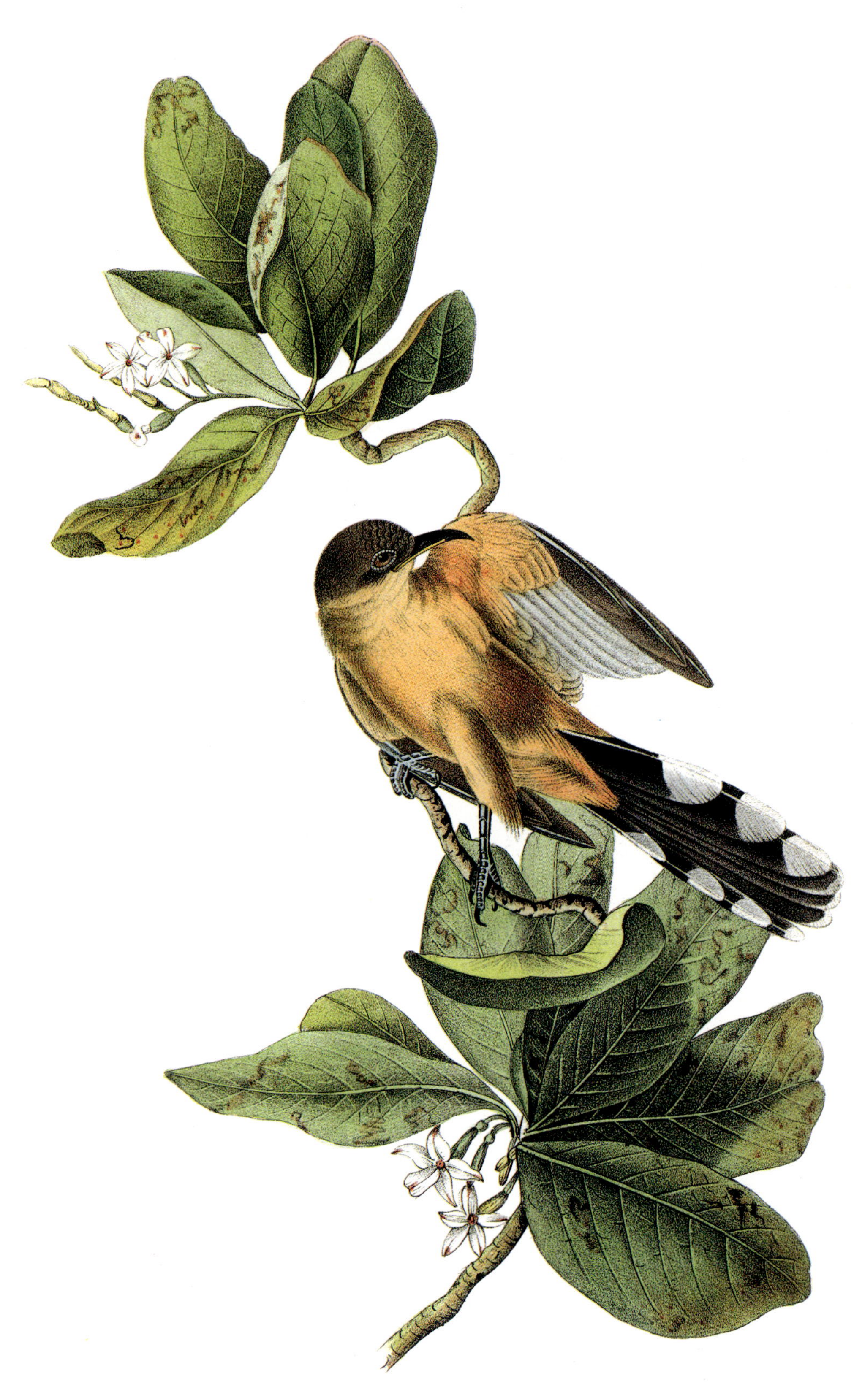

64. Mangrove Cuckoo

Male

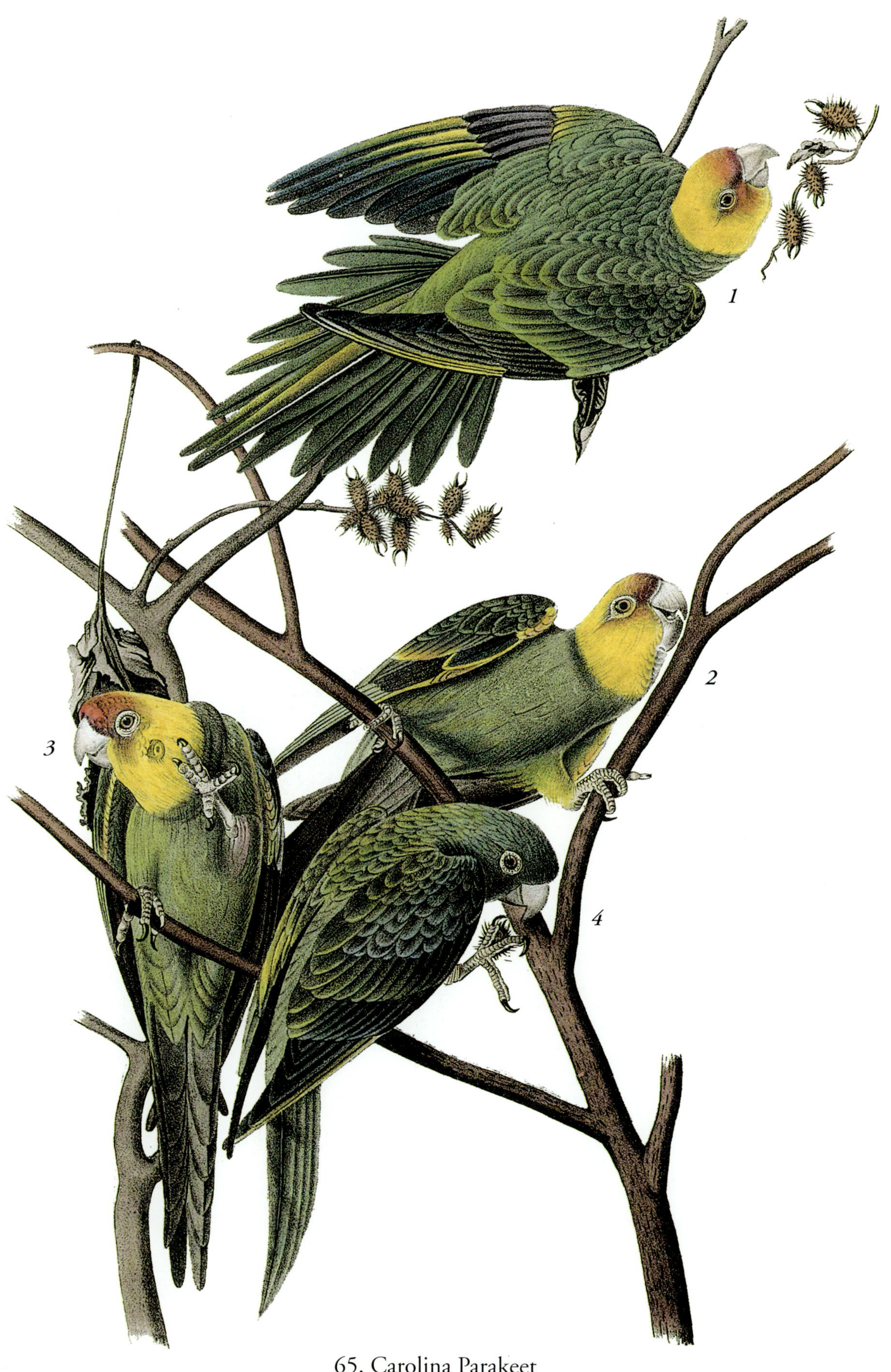

65. Carolina Parakeet

1, 2. Males 3. Female 4. Young

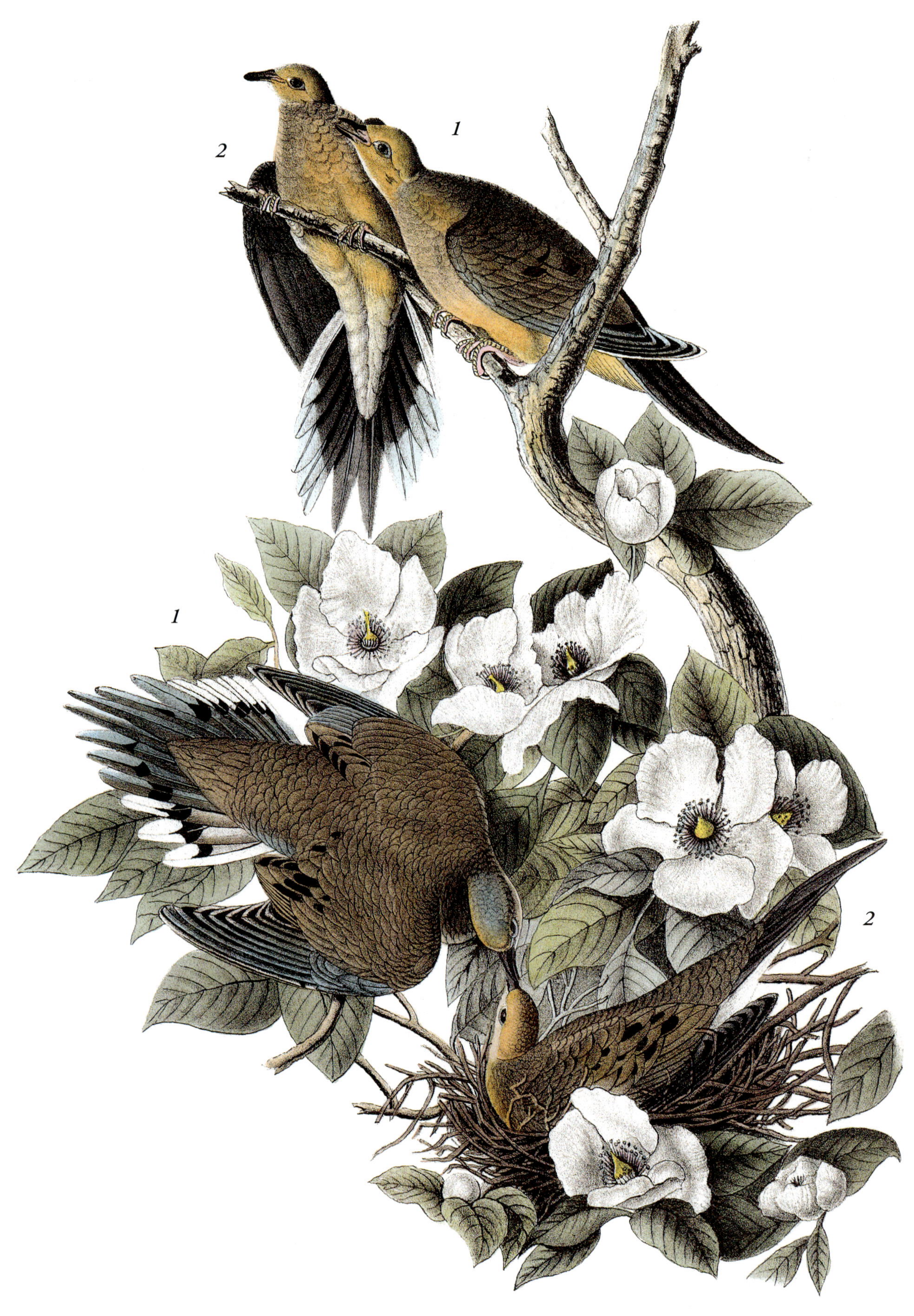

66. Mourning Dove

1. Males 2. Females

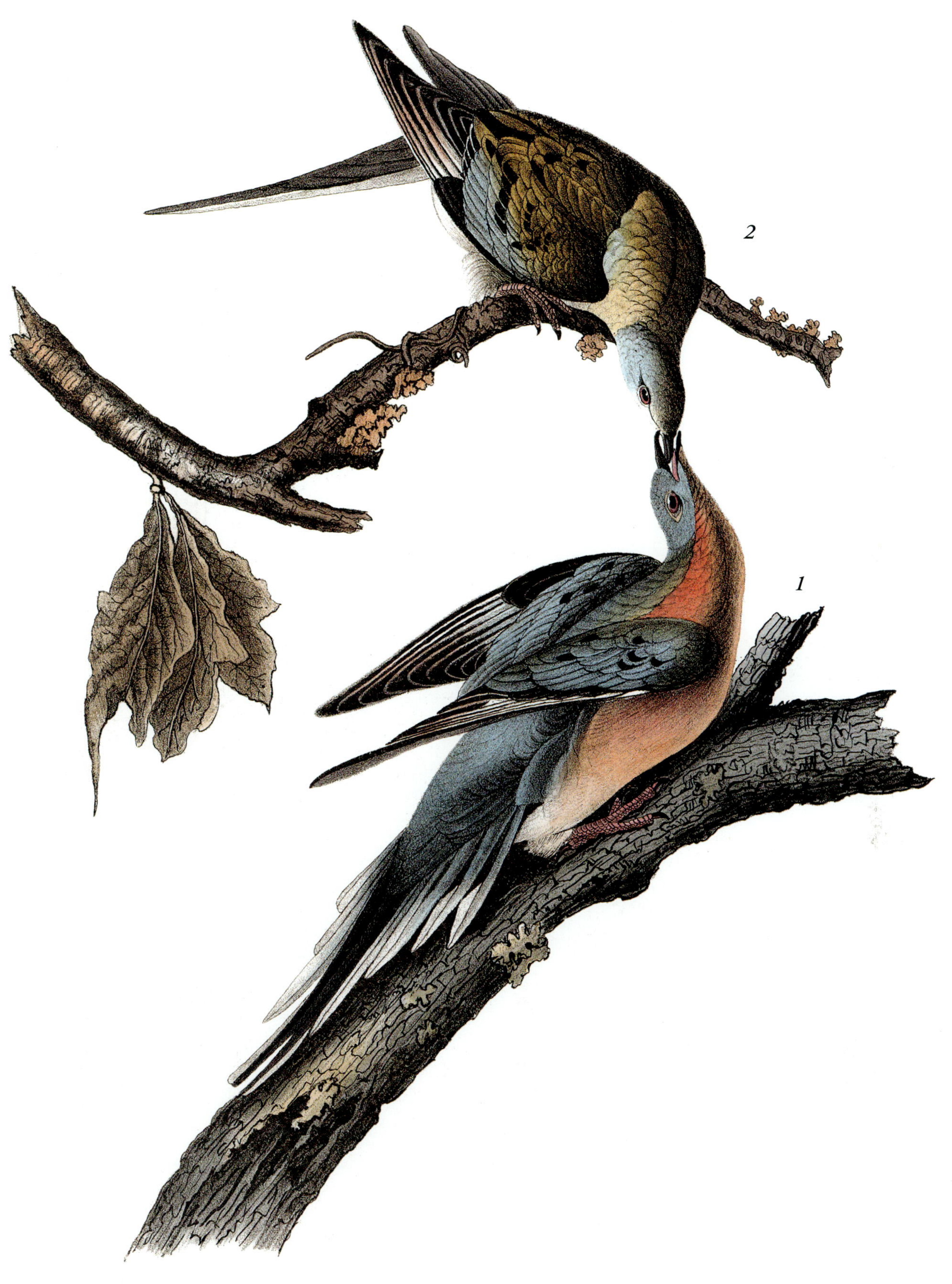

67. Passenger Pigeon

1. Male 2. Female

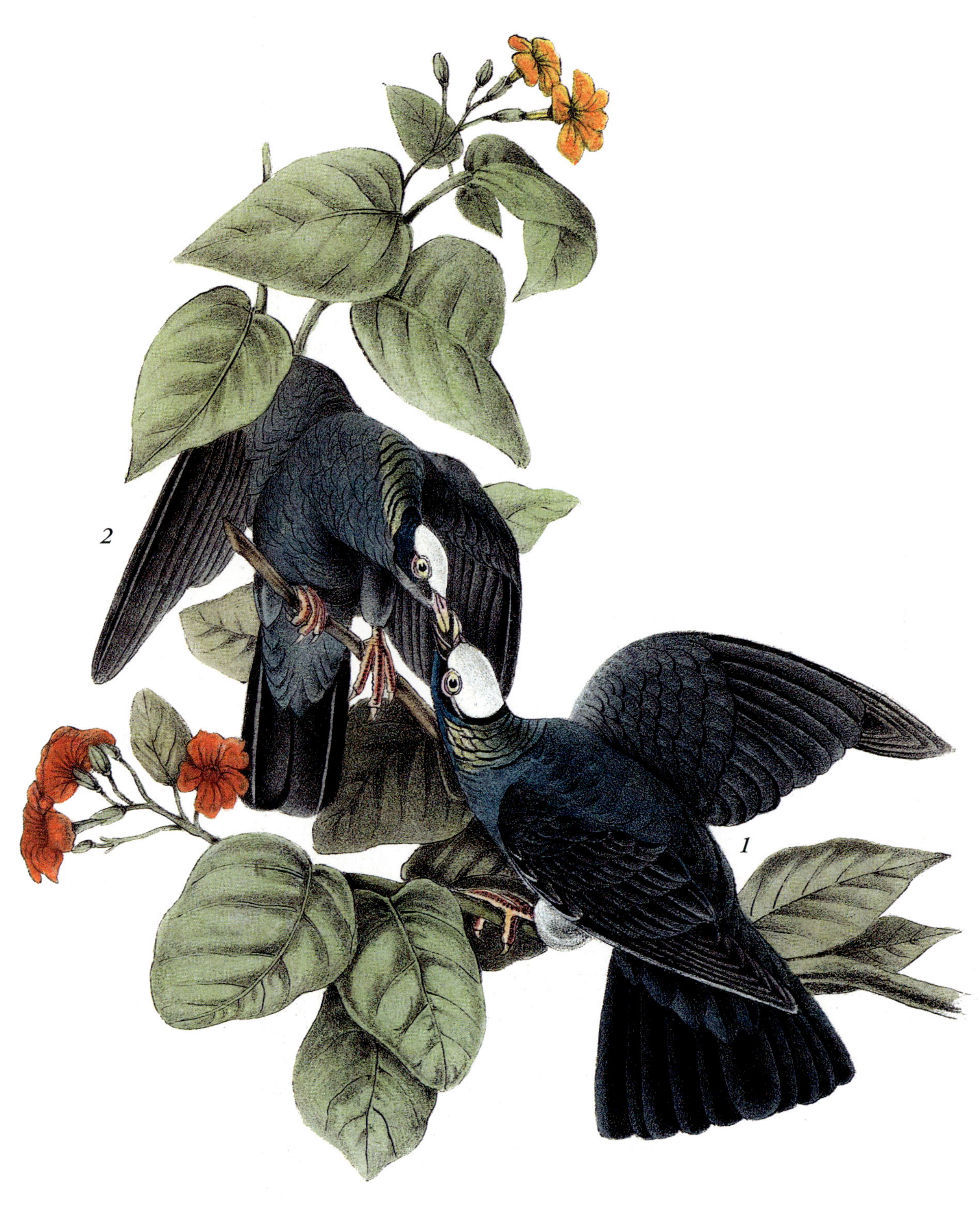

68. White-Crowned Pigeon

1. Male 2. Female

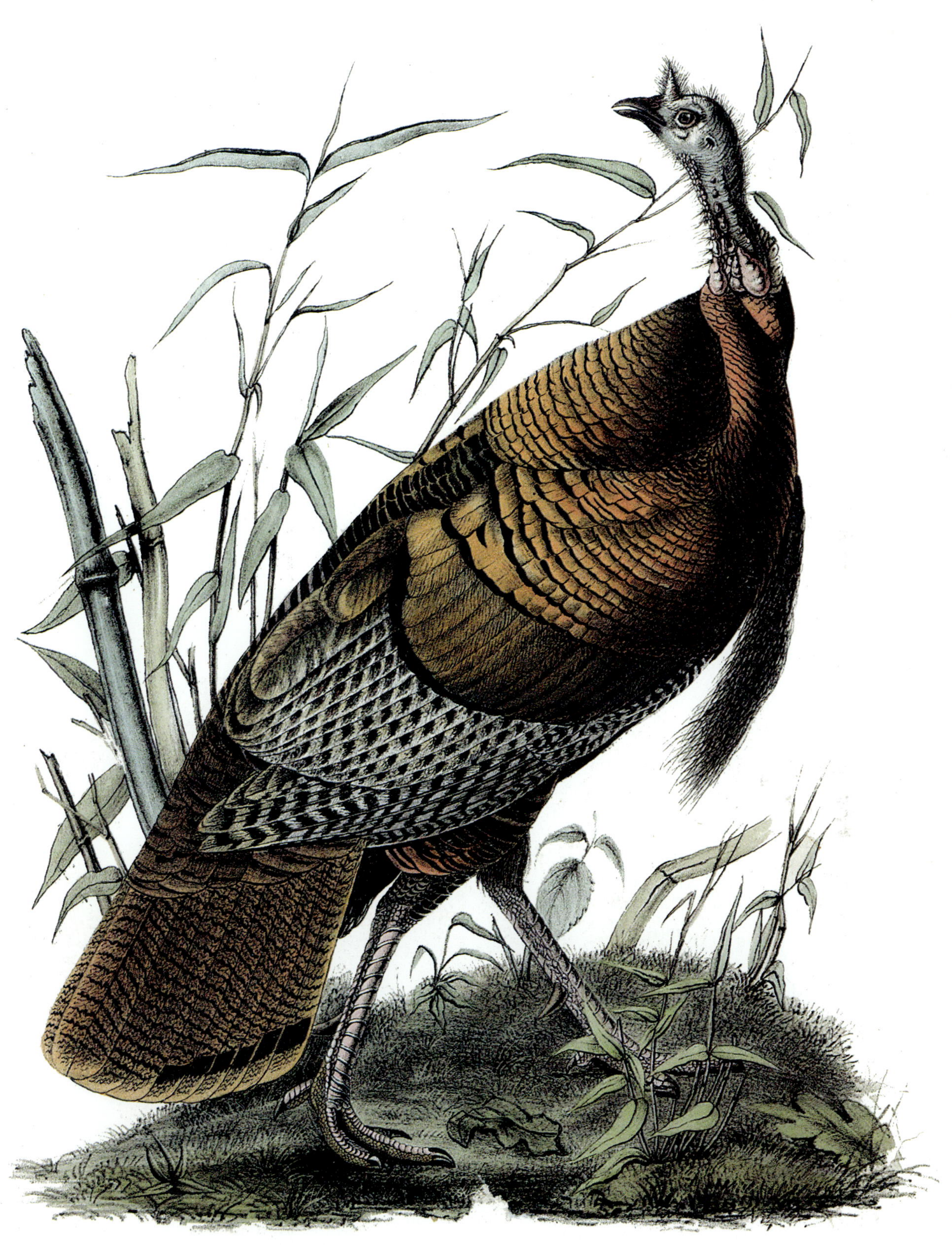

69. Wild Turkey

Male

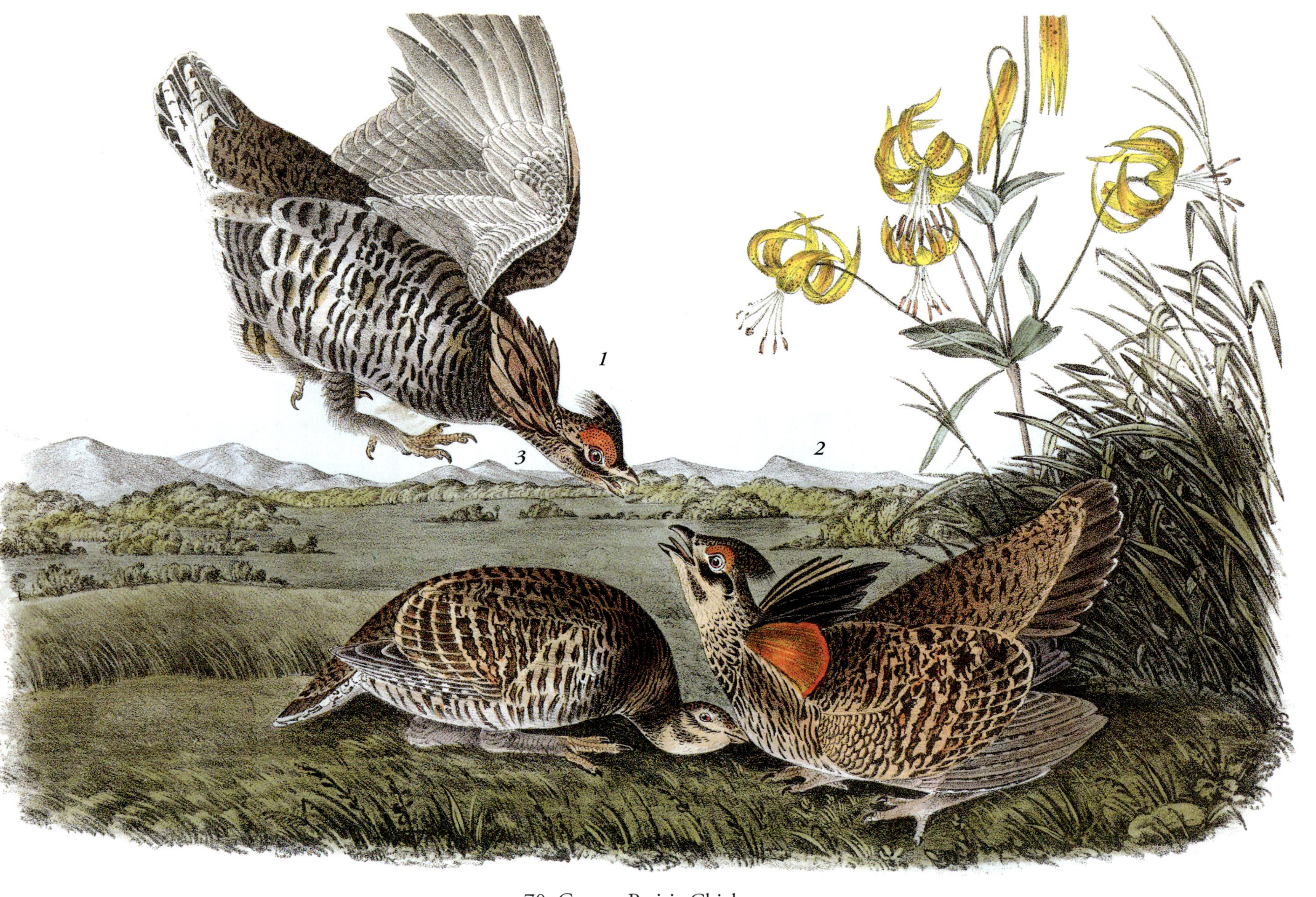

70. Greater Prairie Chicken

1, 2. Males 3. Female

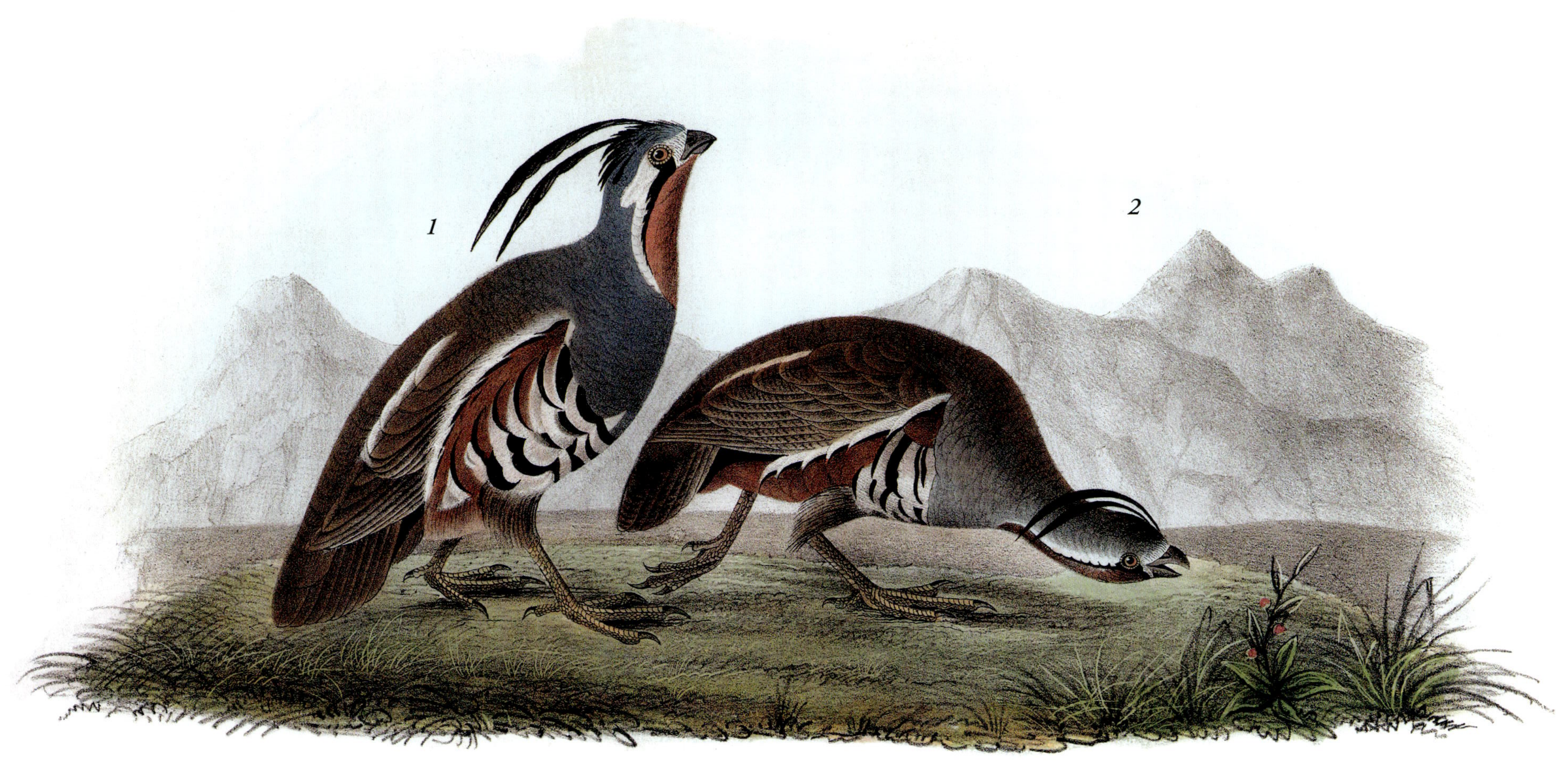

71. Mountain Quail

1. Male 2. Female

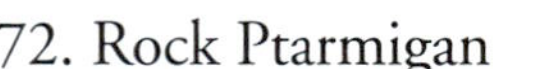

72. Rock Ptarmigan

1. Male in Winter 2. Female, Summer Plumage 3. Young in August

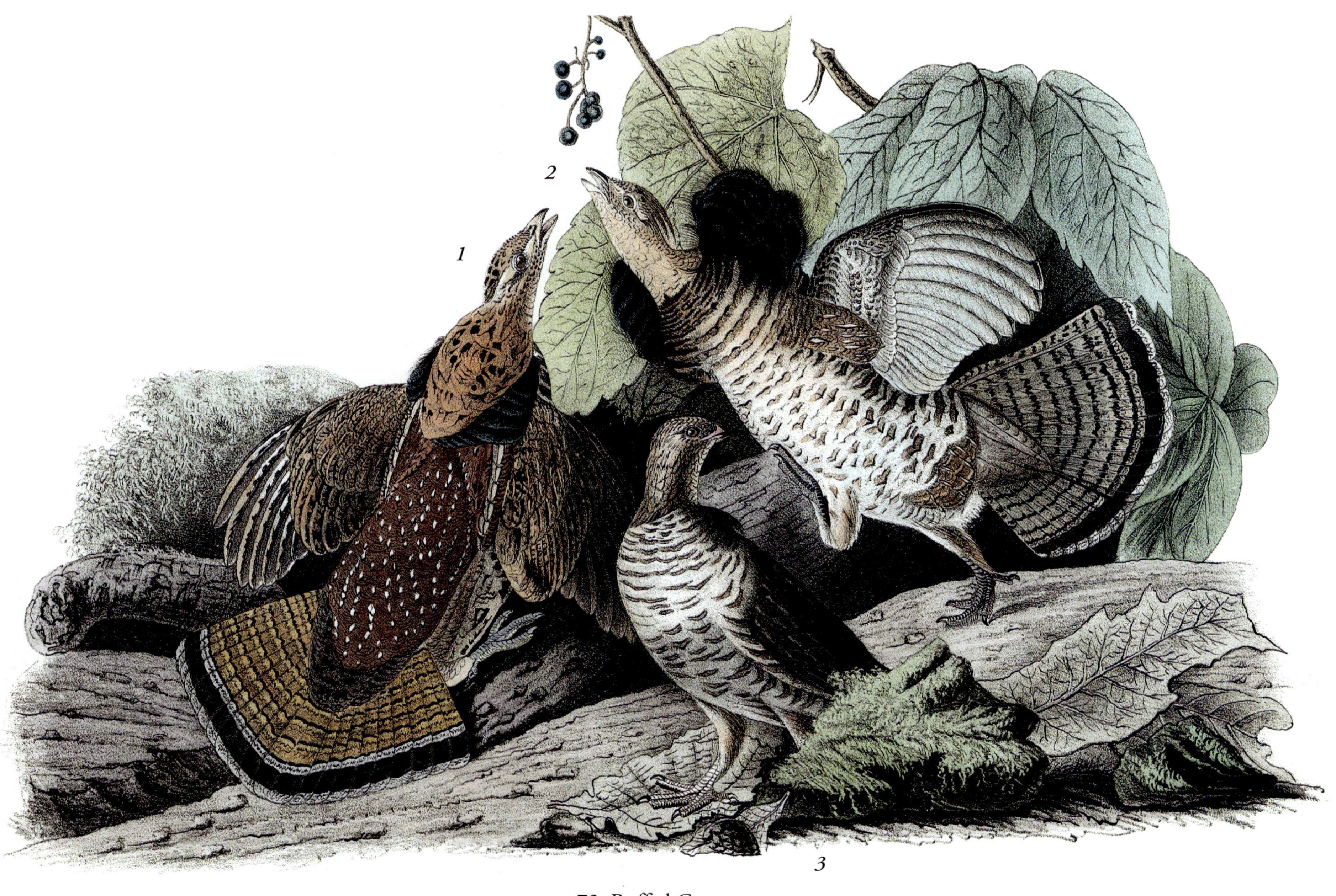

73. Ruffed Grouse

1, 2. Males 3. Female

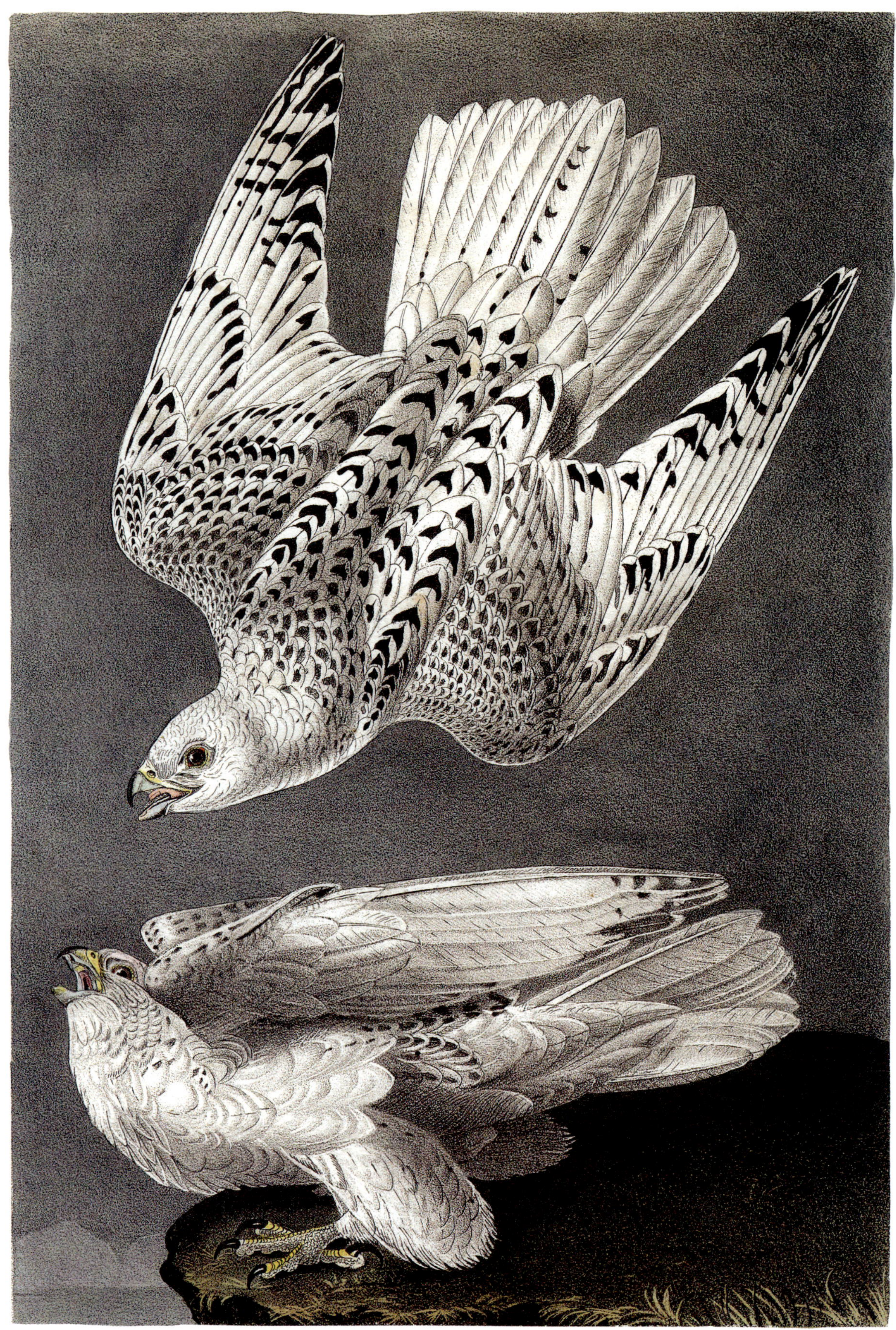

74. Gyrfalcon

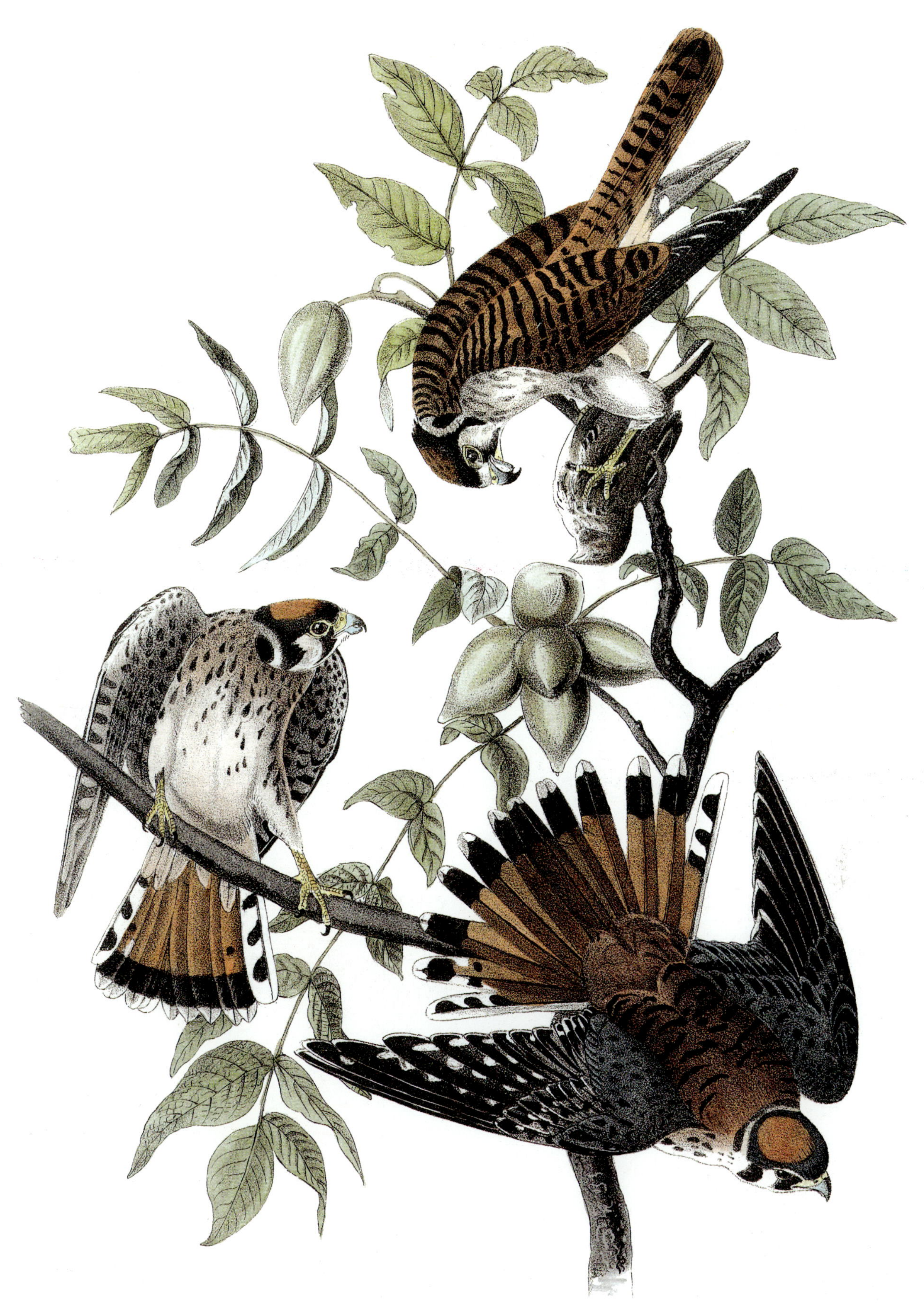

75. American Kestrel

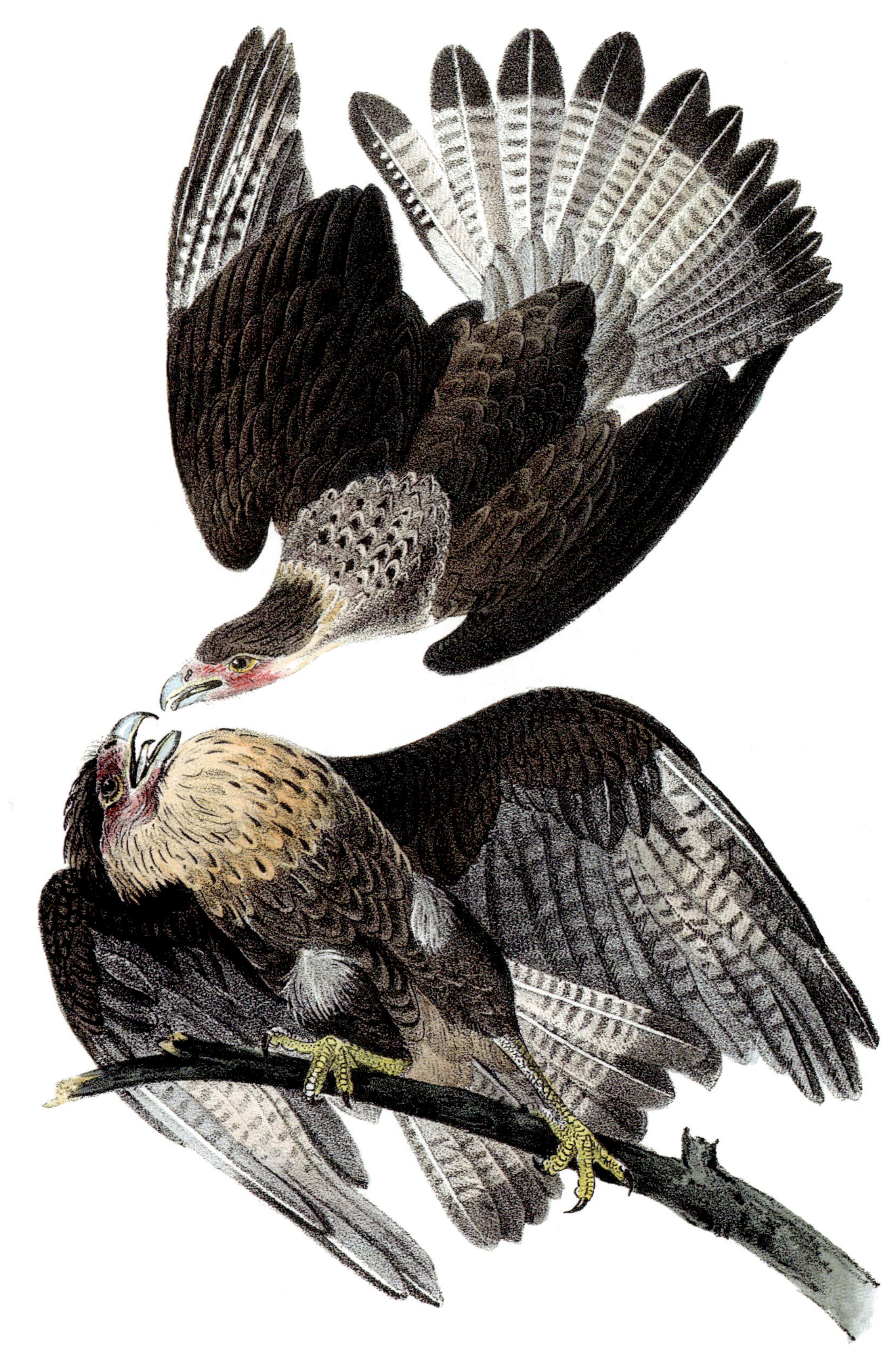

76. Crested Caracara

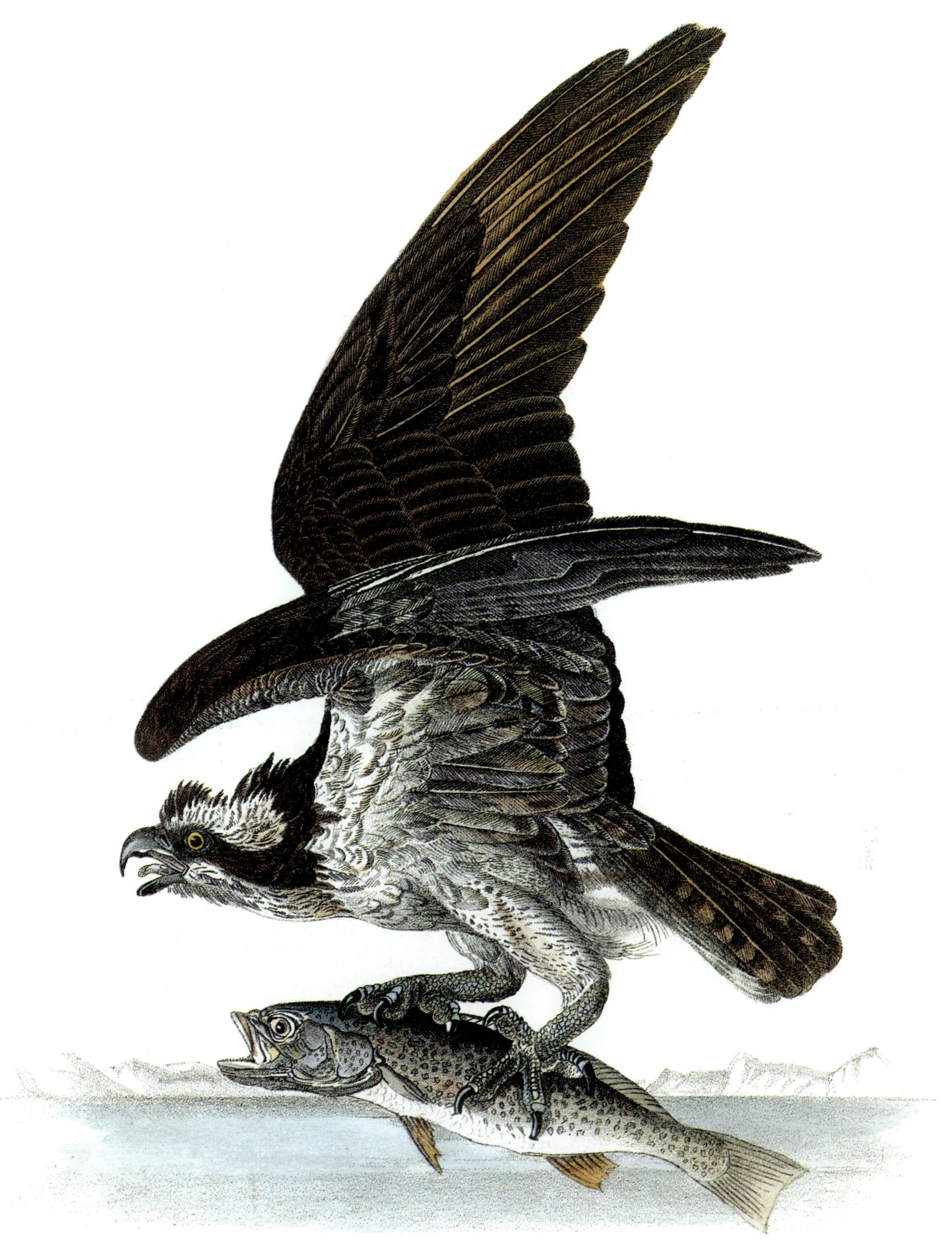

77. Osprey

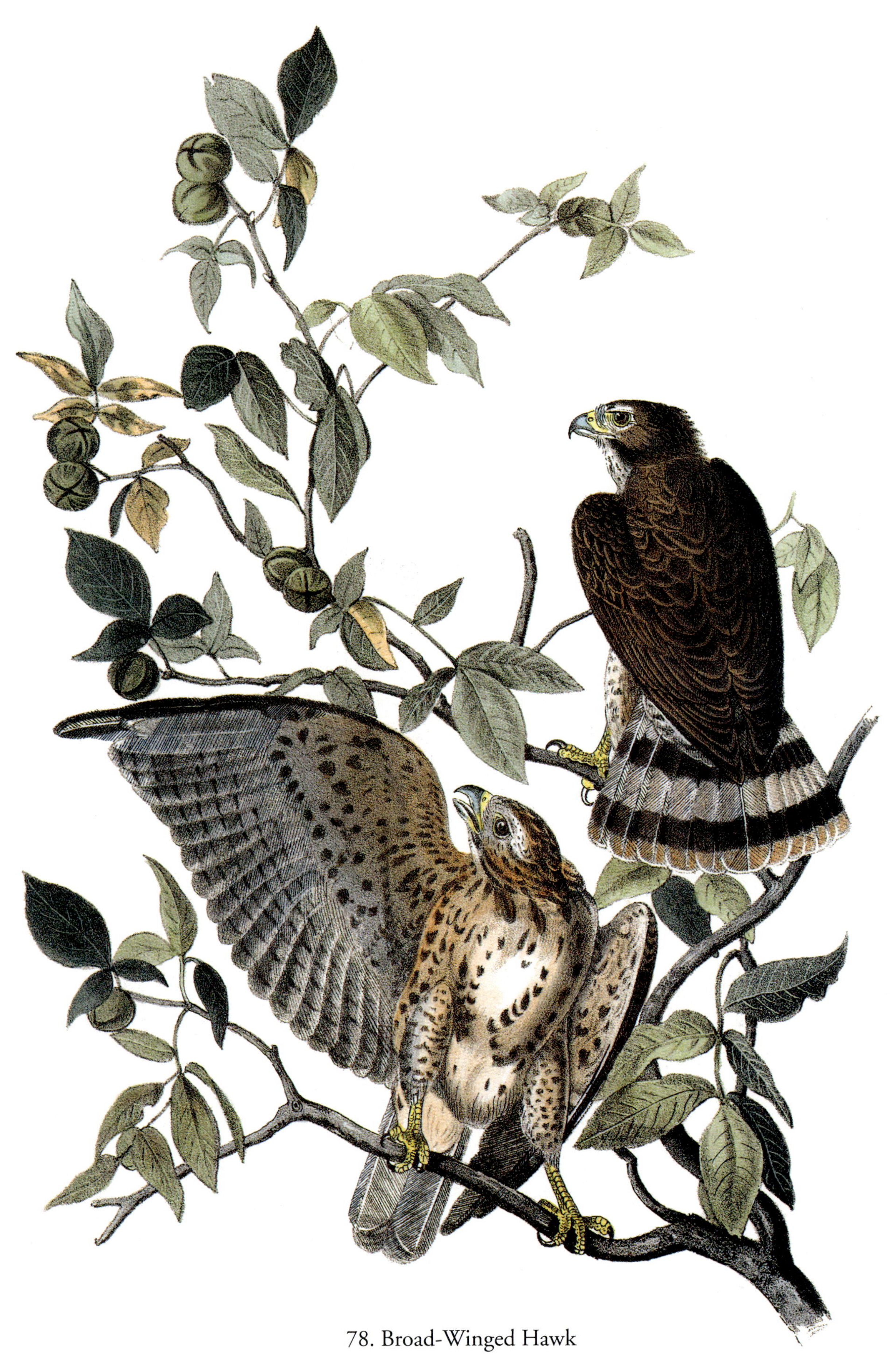

78. Broad-Winged Hawk

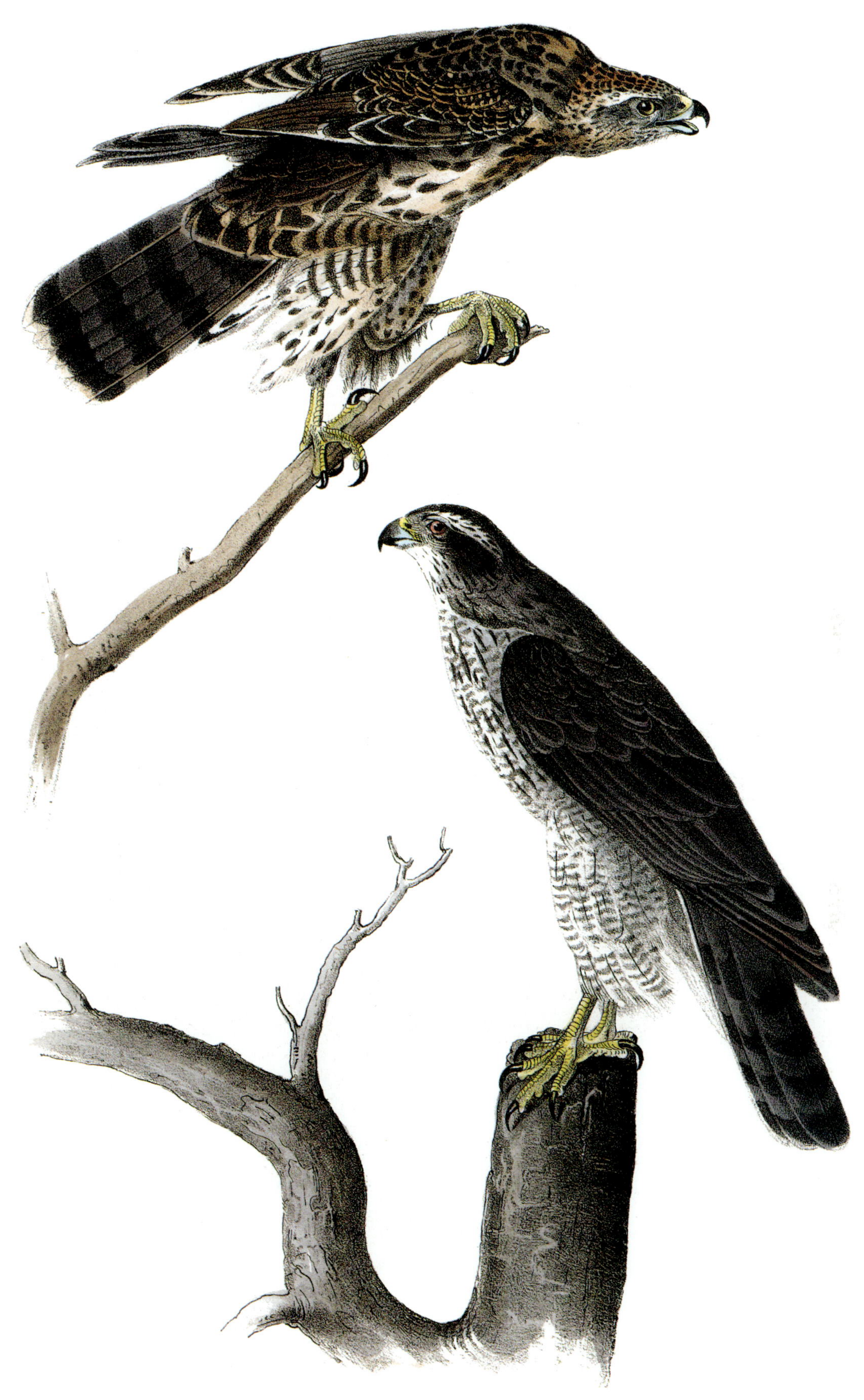

79. Northern Goshawk

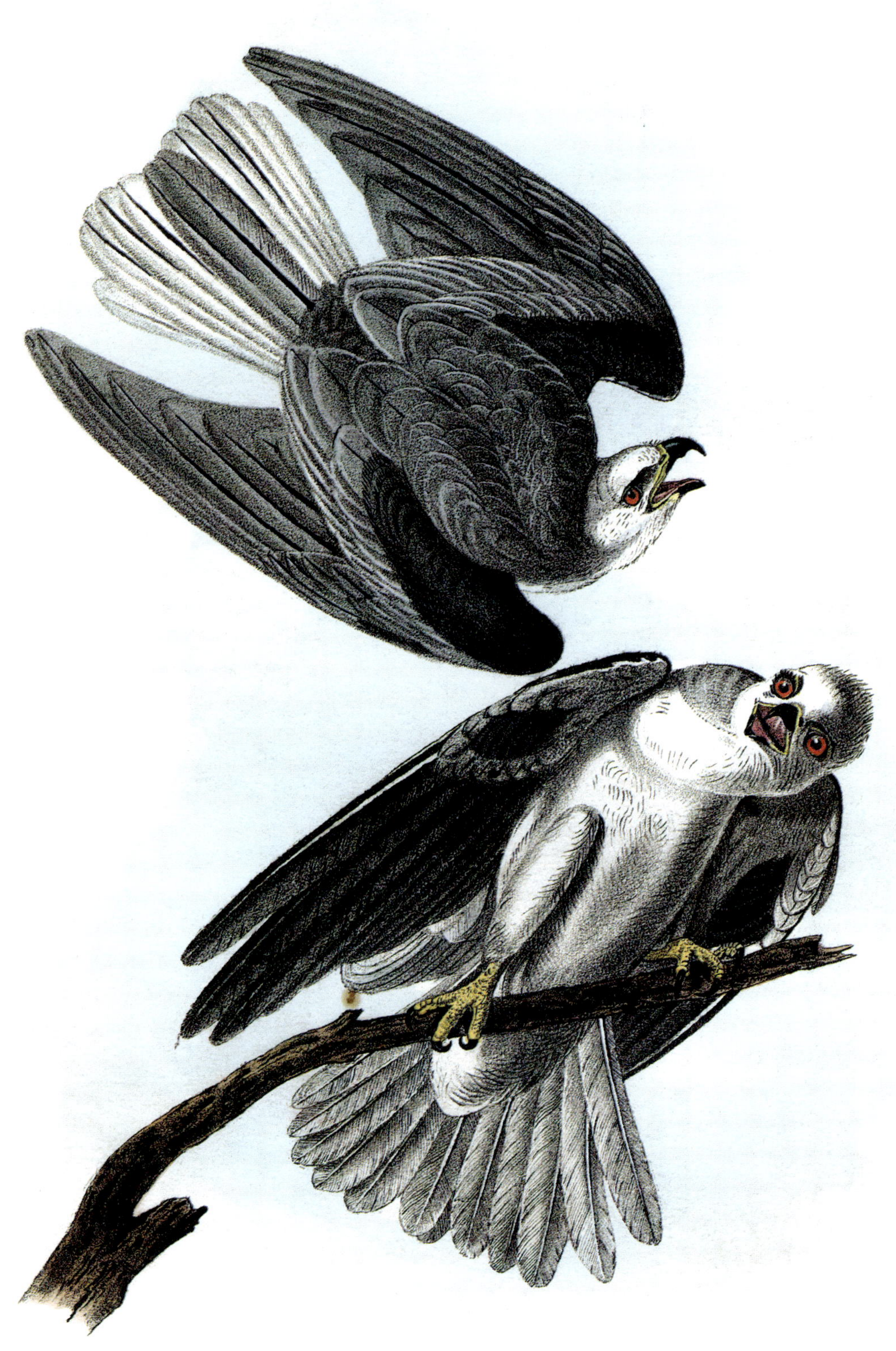

80. Black-Shouldered Kite

81. Black Vulture

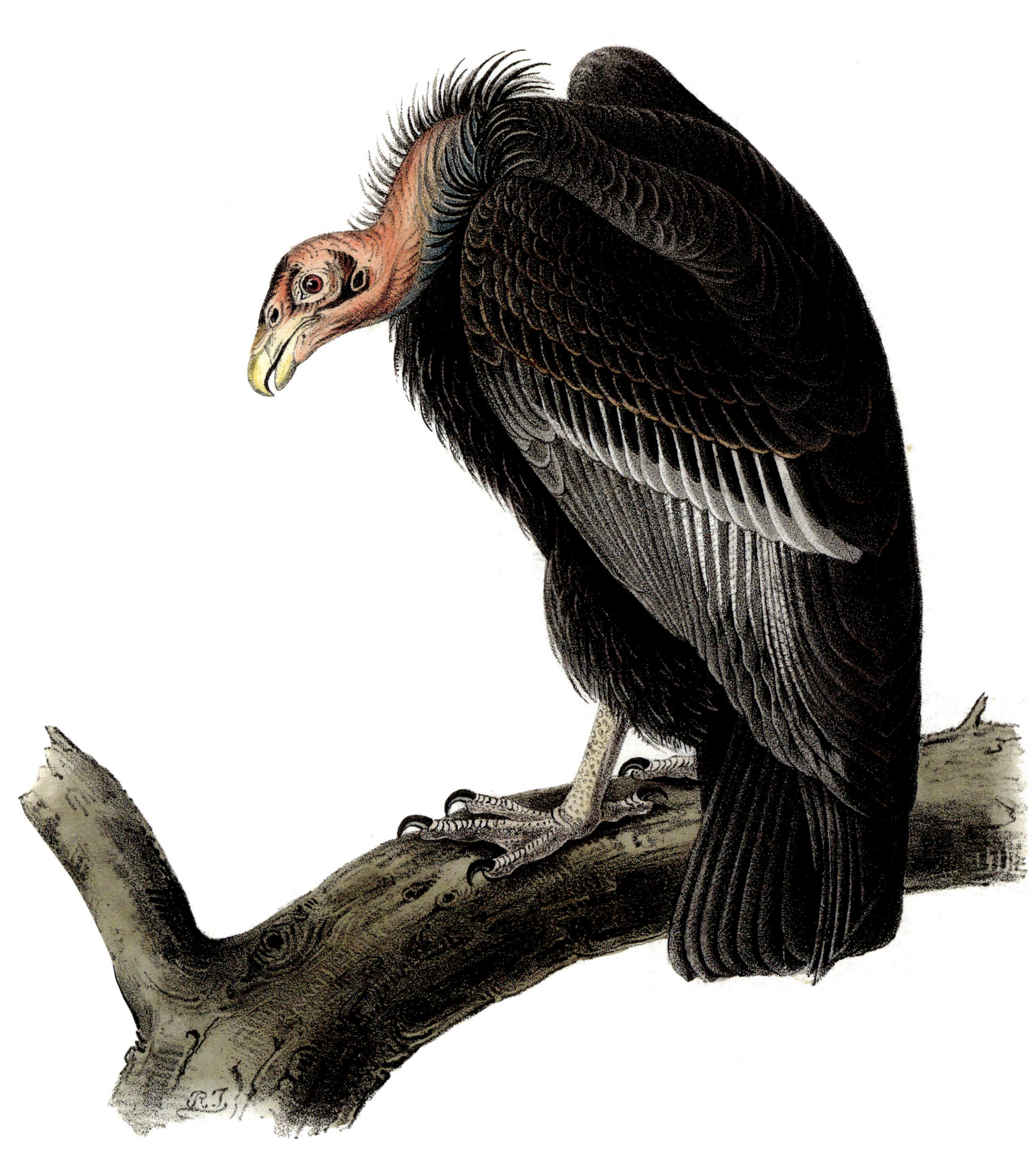

82. California Condor

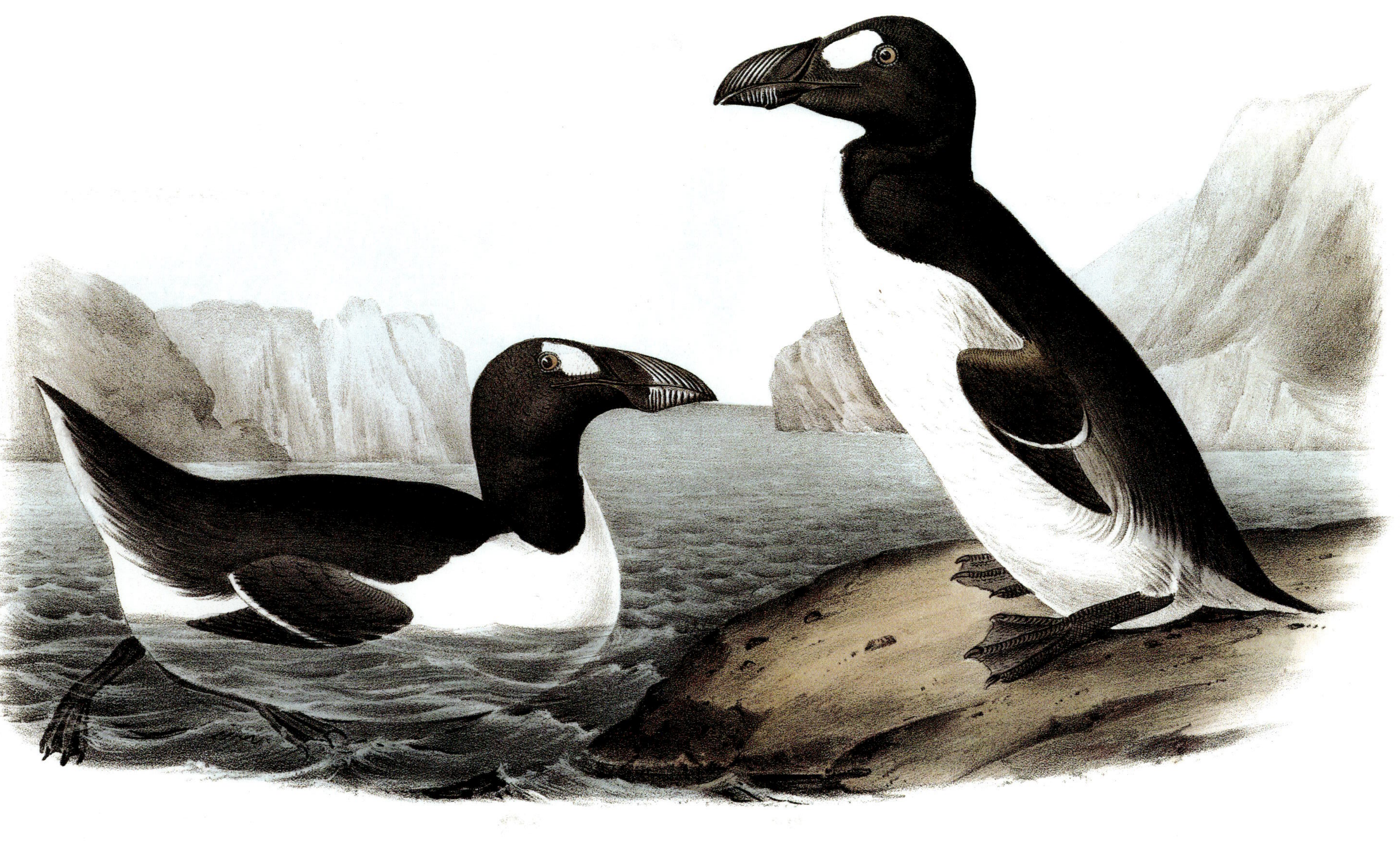

83. Great Auk

Adult

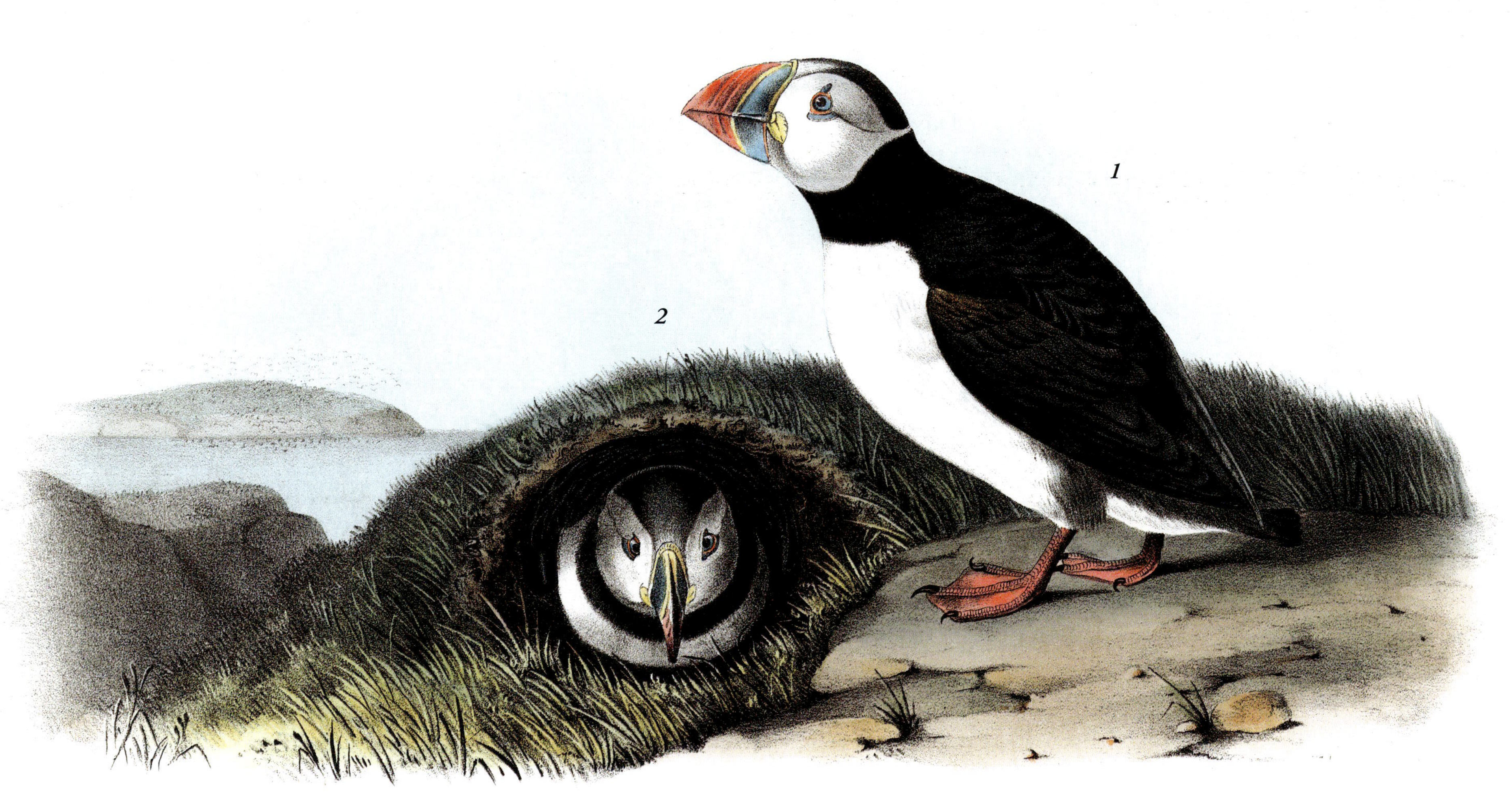

84. Atlantic Puffin

1. Male 2. Female

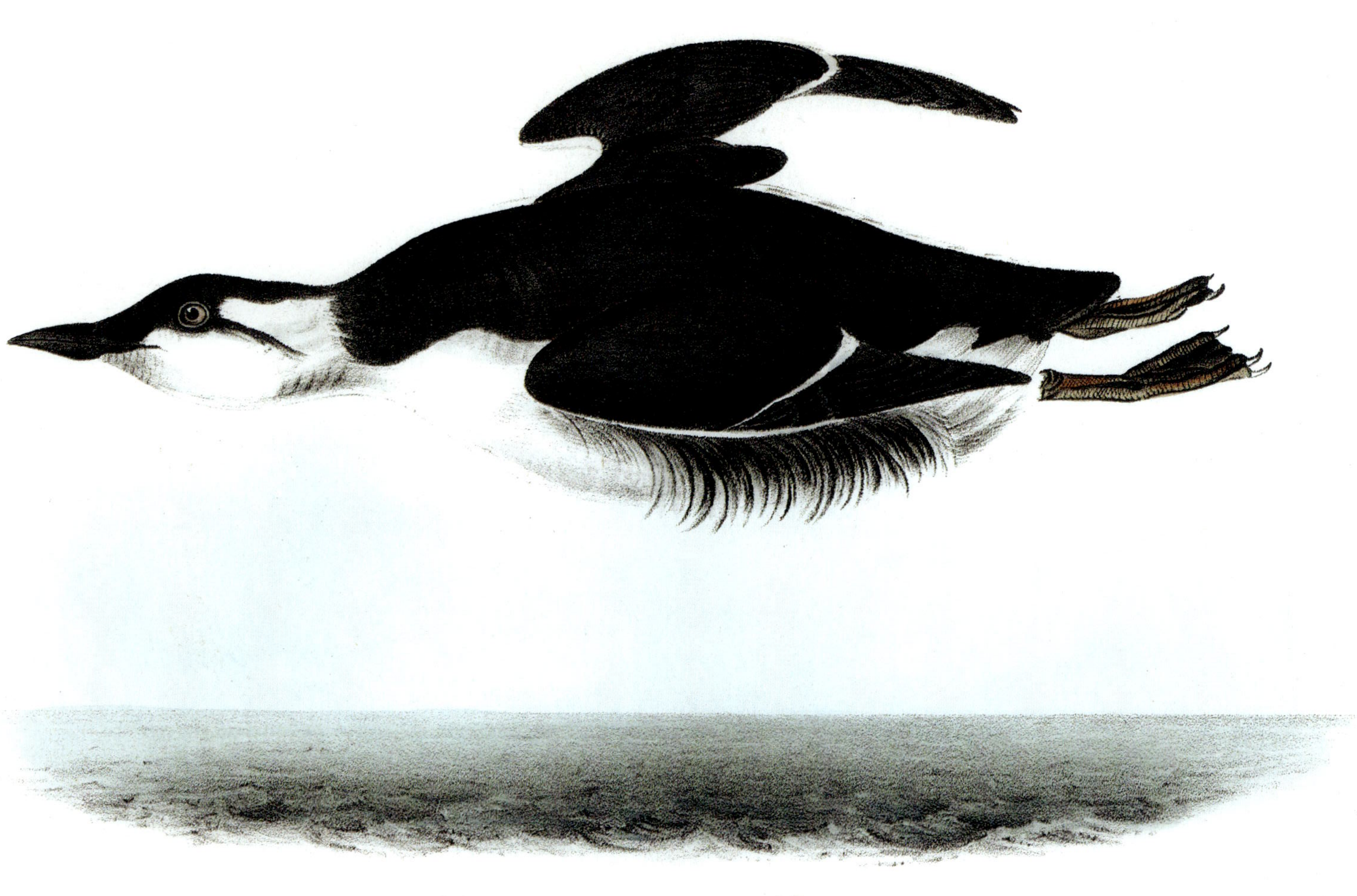

85. Thick-Billed Murre

Male

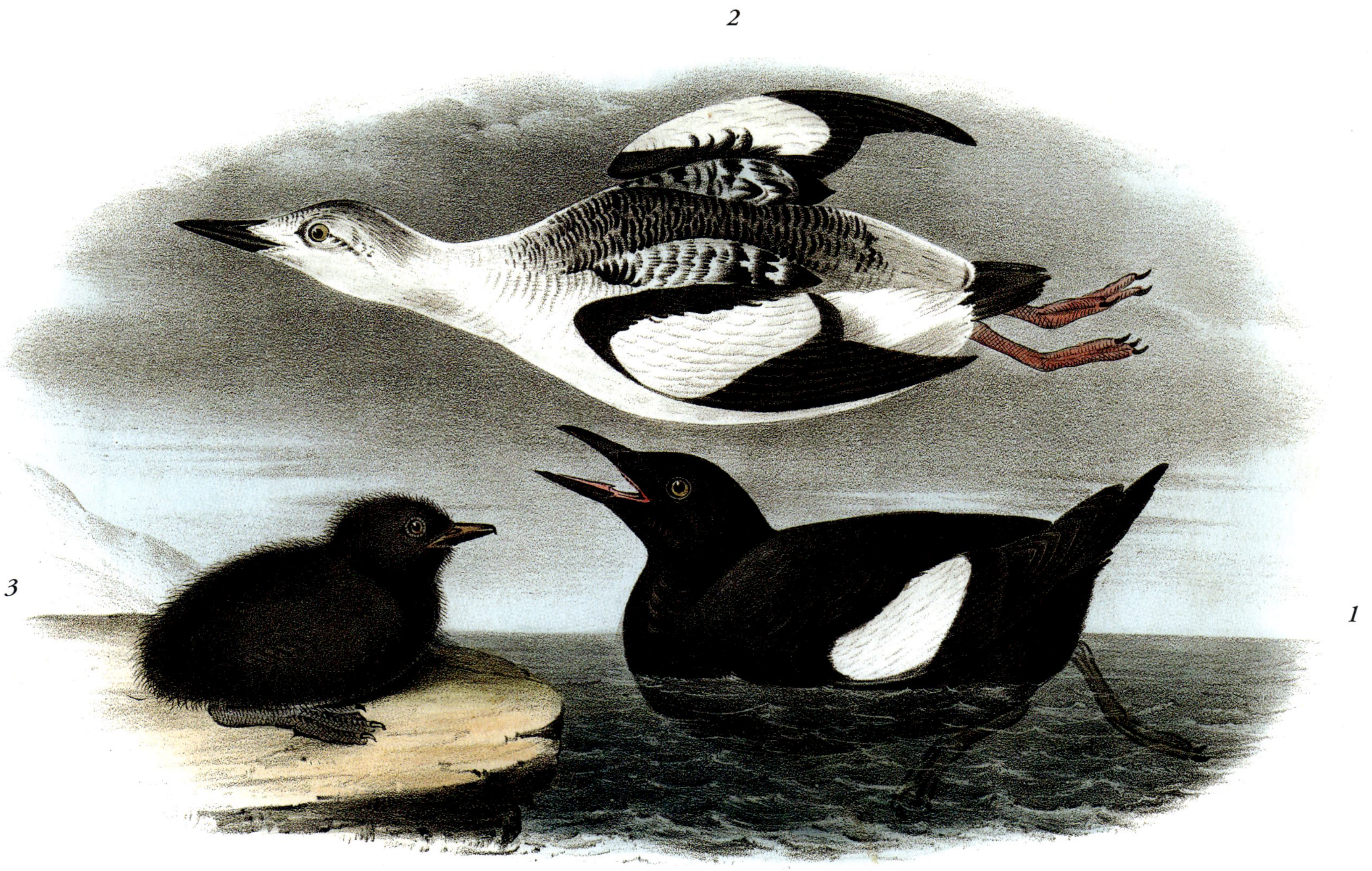

2

3

1

86. Black Guillemot

1. Adult, Summer Plumage   2. Adult in Winter   3. Young

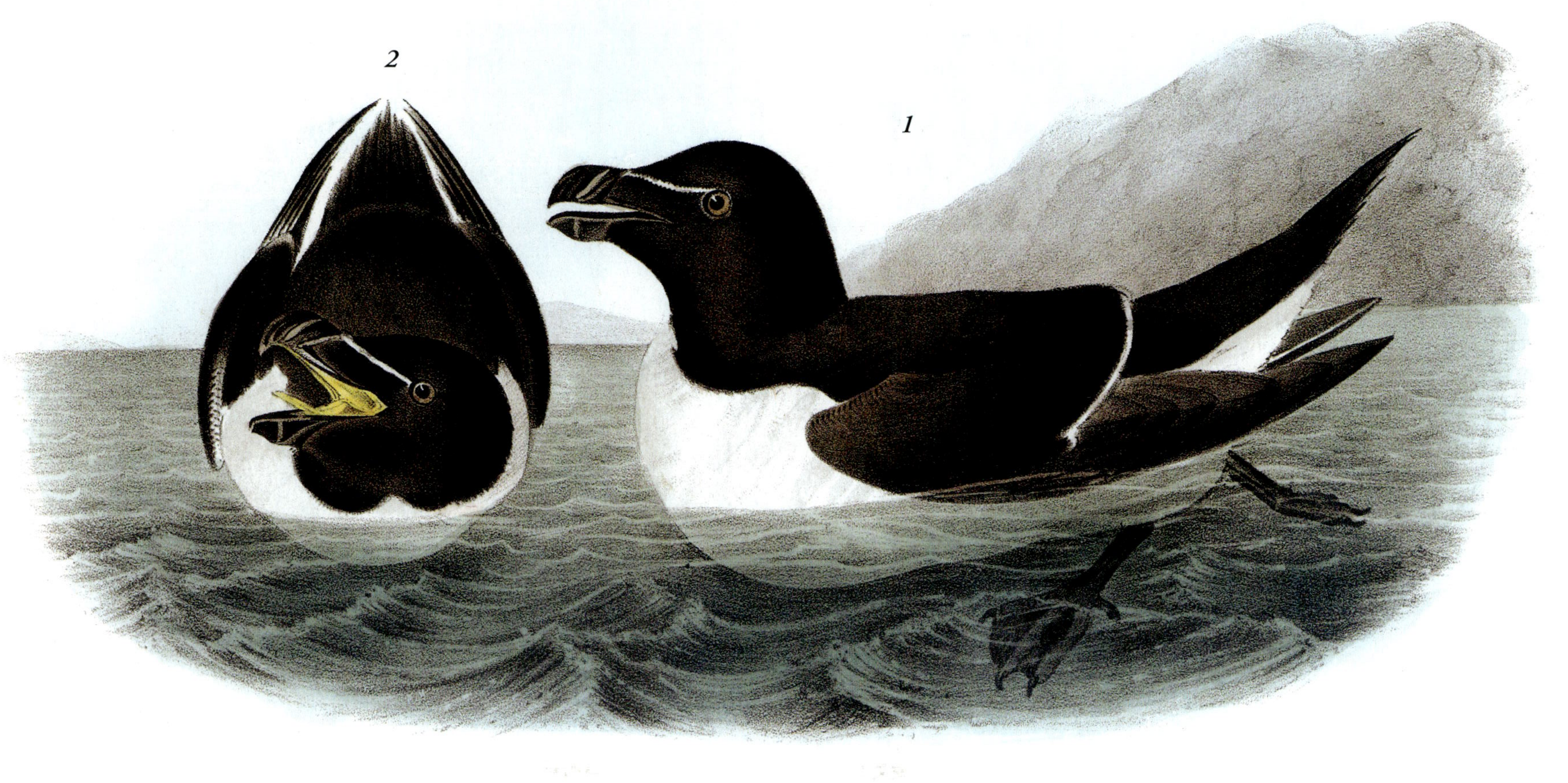

87. Razorbill

1. Male 2. Female

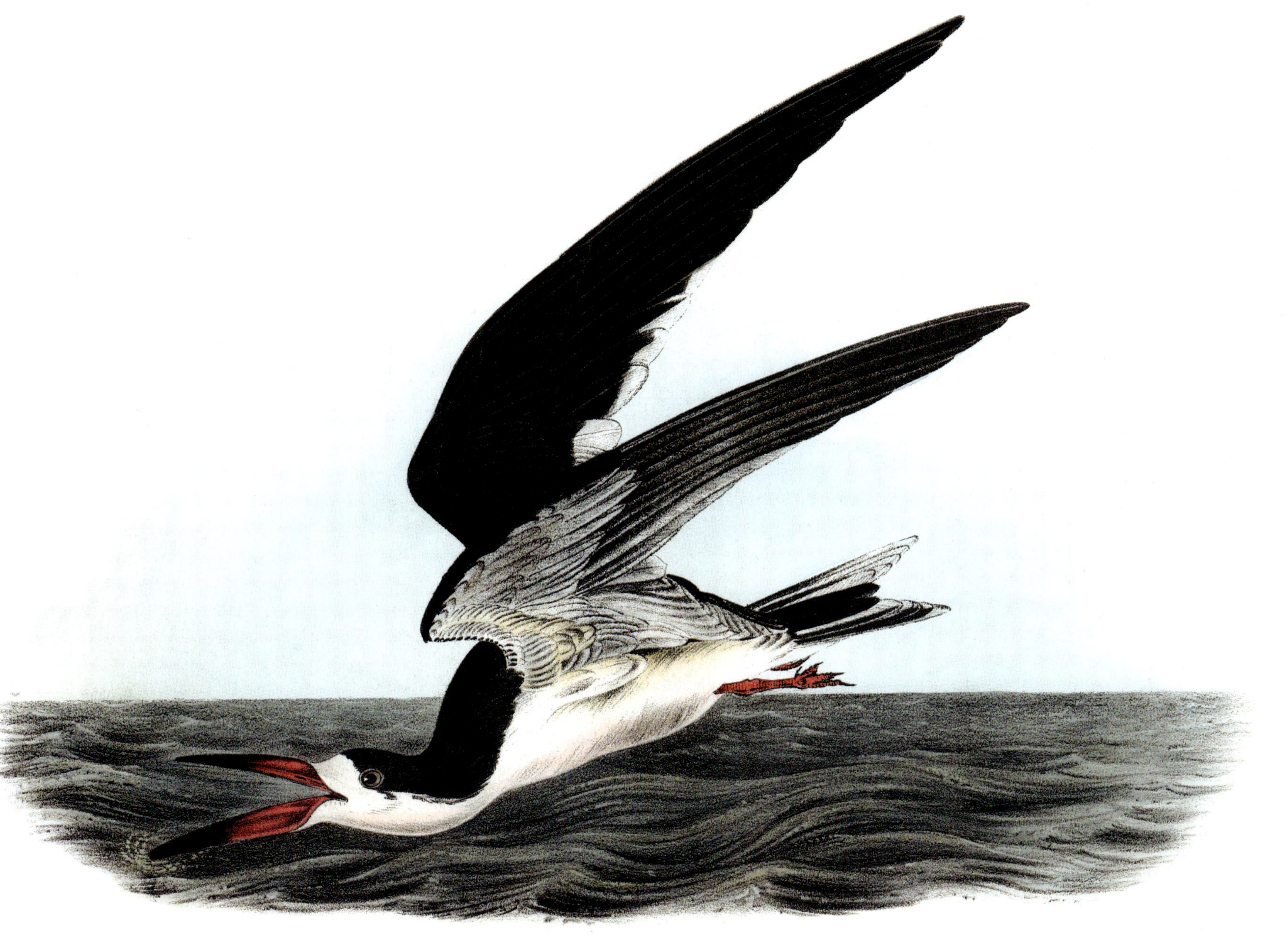

88. Black Skimmer

Male

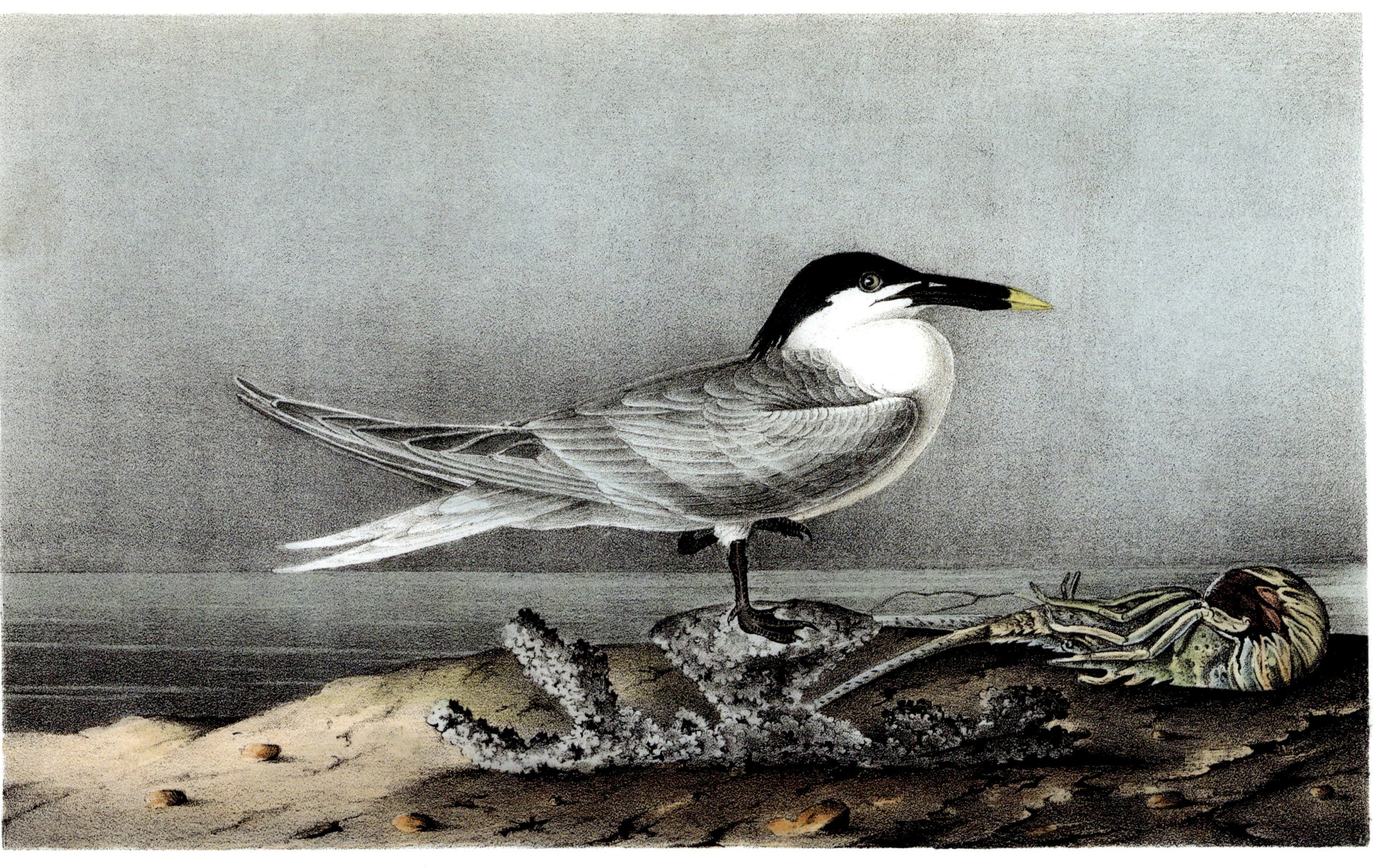

89. Sandwich Tern

Adult

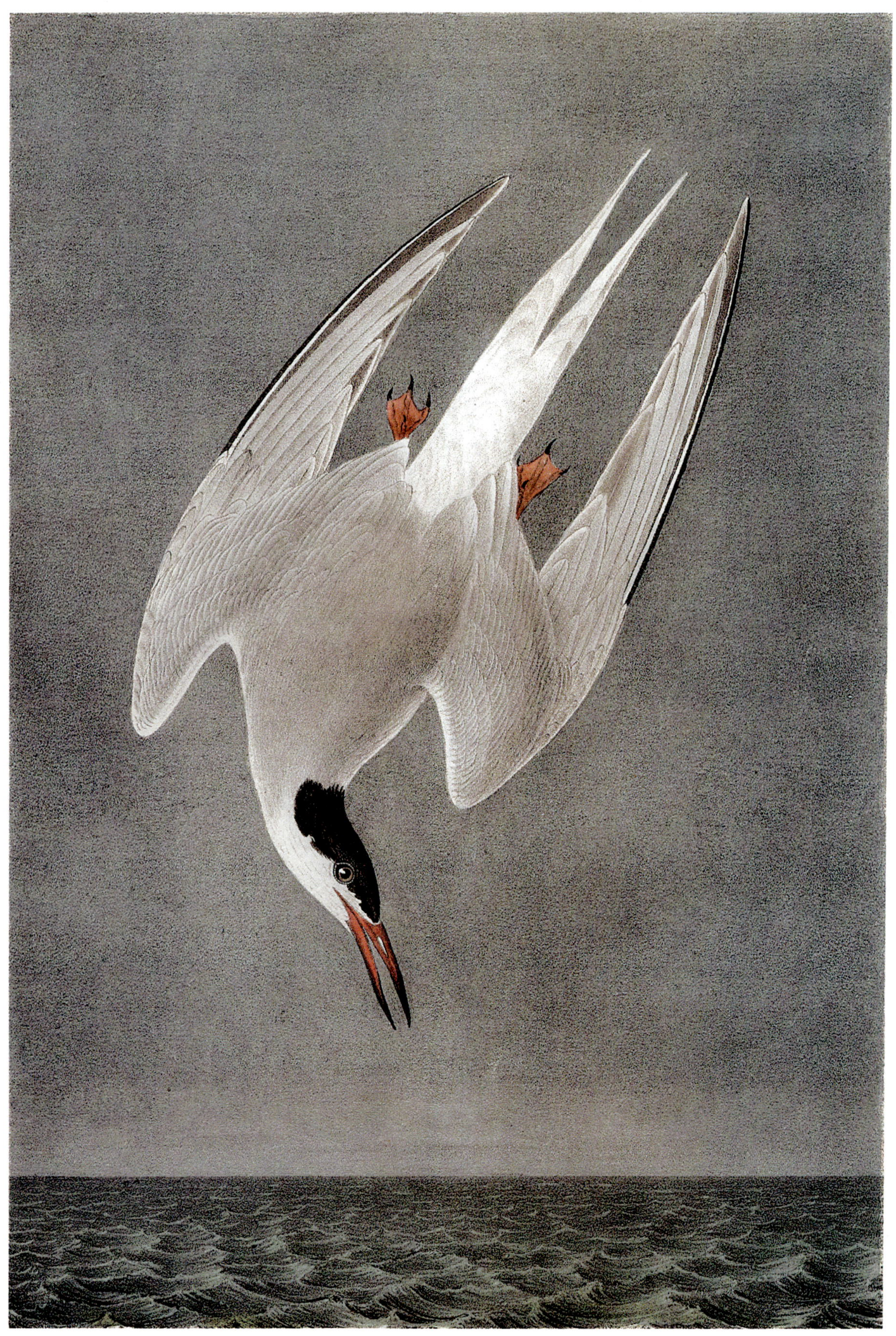

90. Arctic Tern

Male

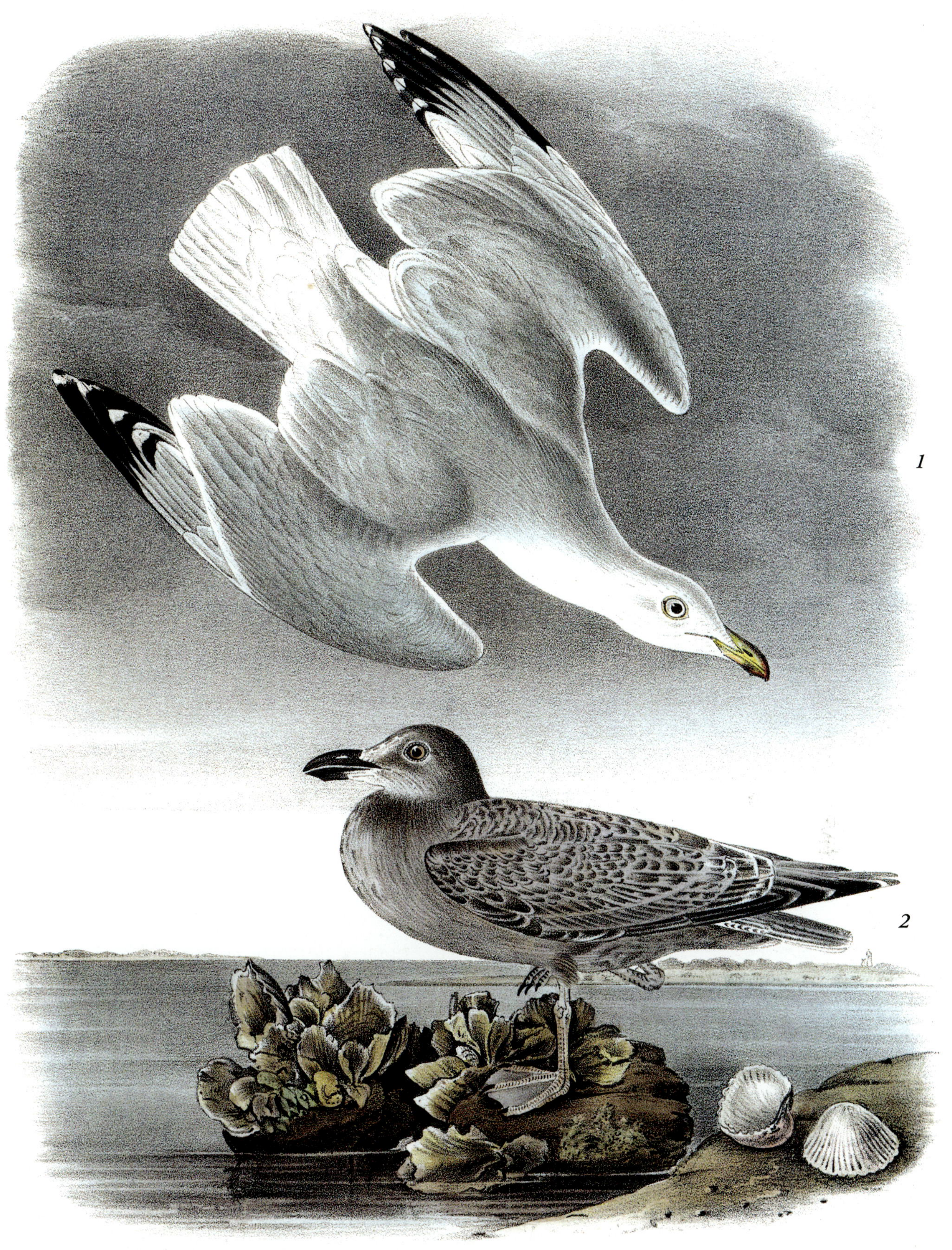

91. Herring Gull

1. Adult in Spring 2. Young in Autumn

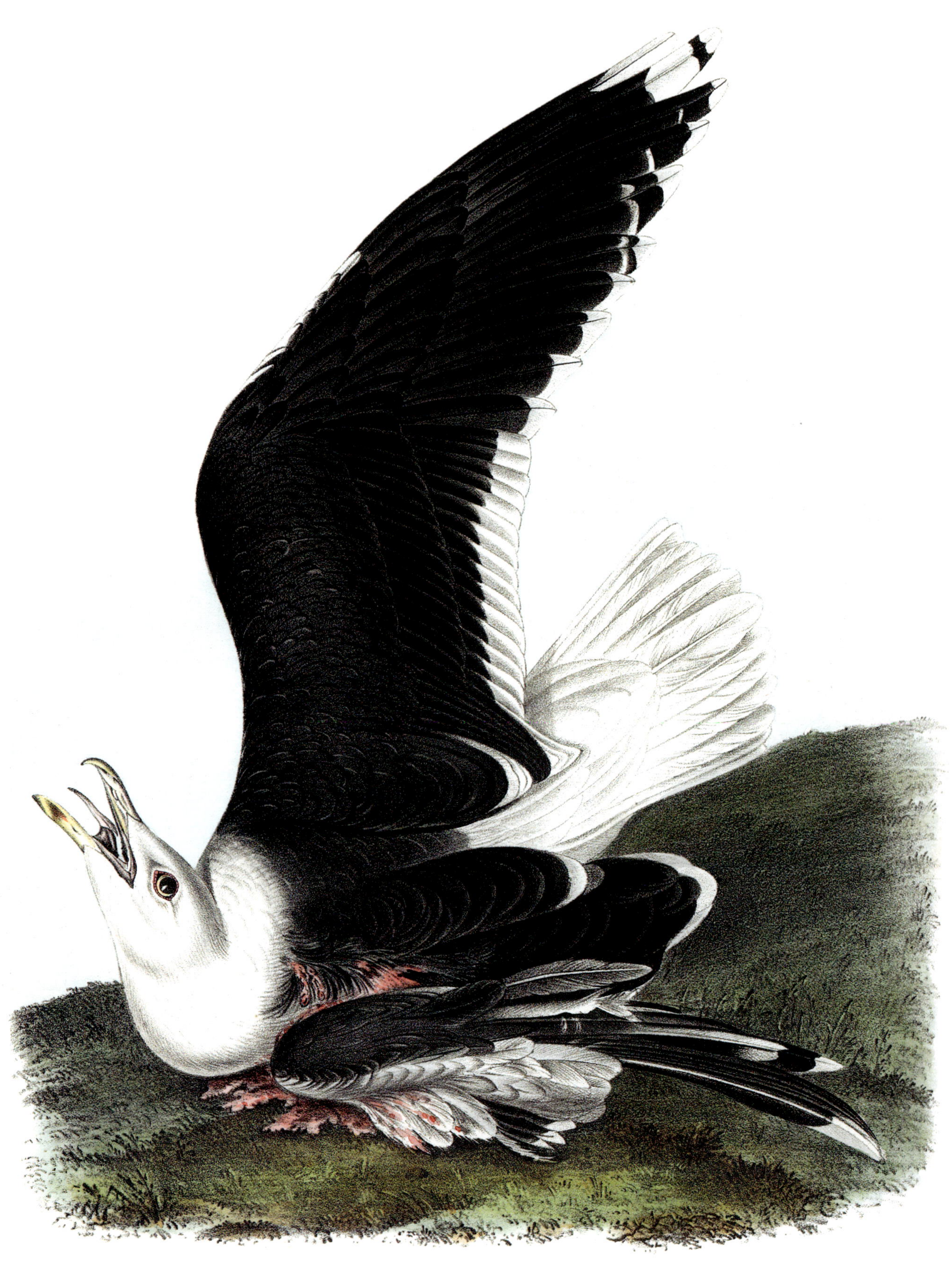

92. Great Black-Backed Gull

Male

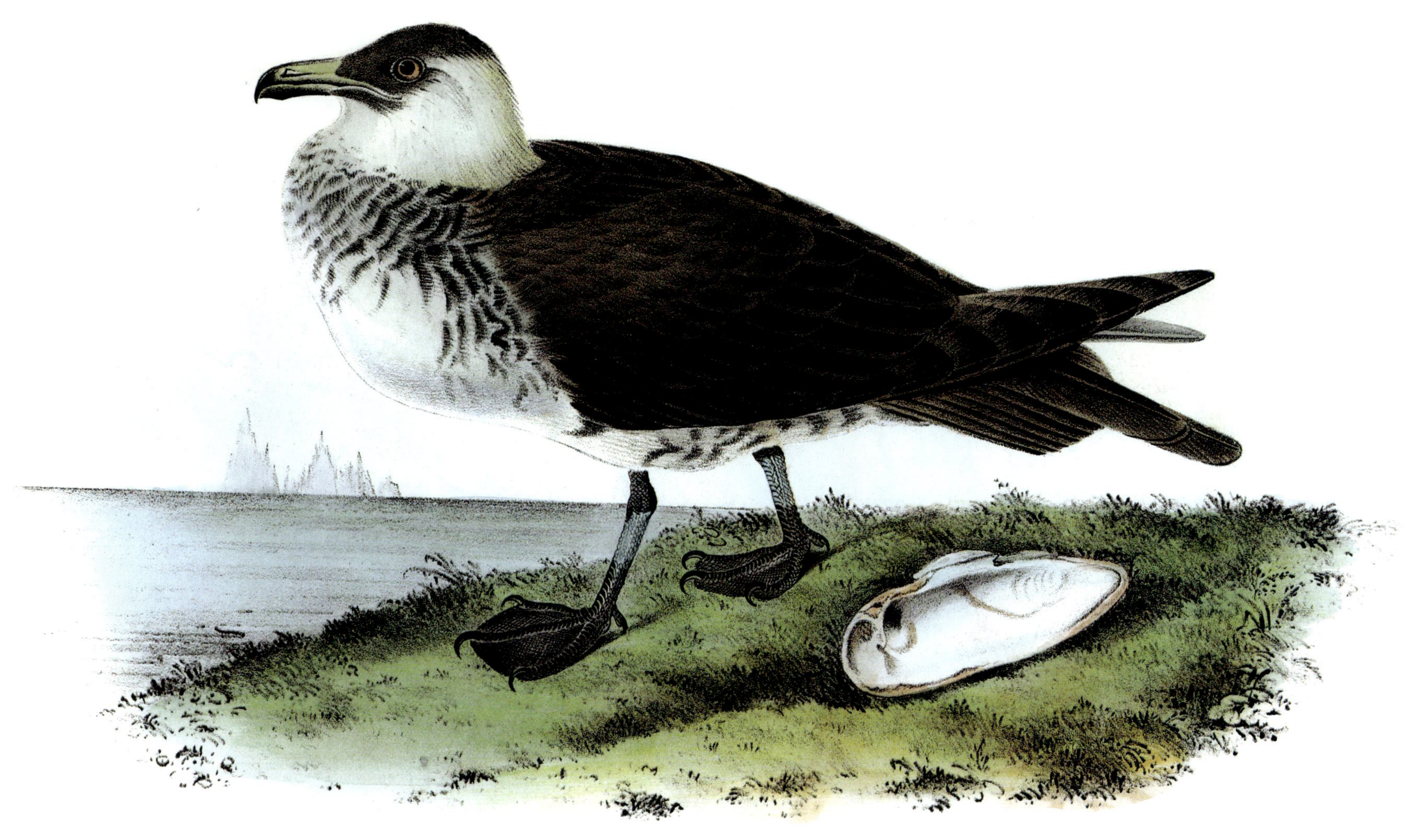

93. Pomarine Jaeger

Adult Female

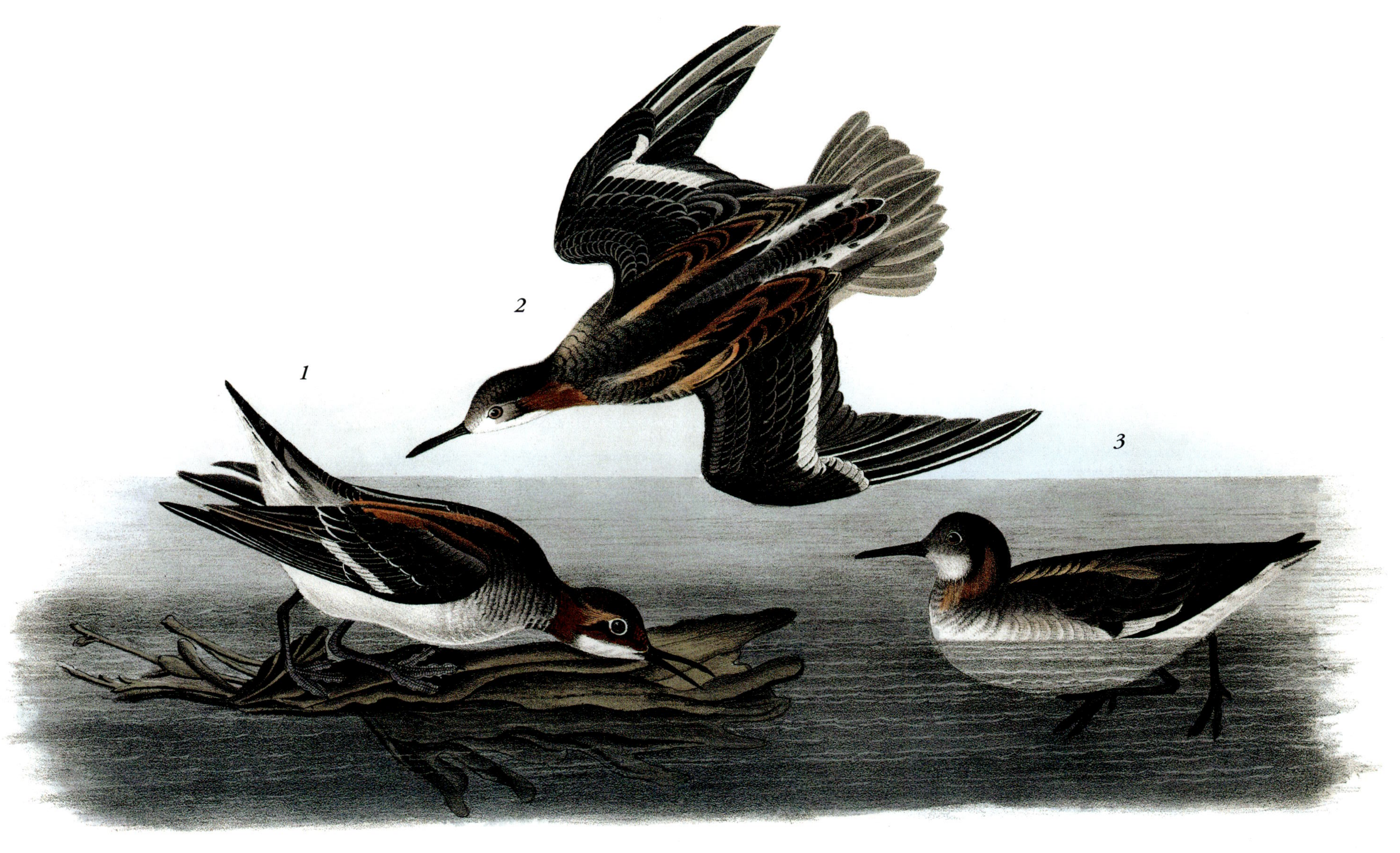

94. Red-Necked Phalarope

1. Male 2. Female 3. Young in Autumn

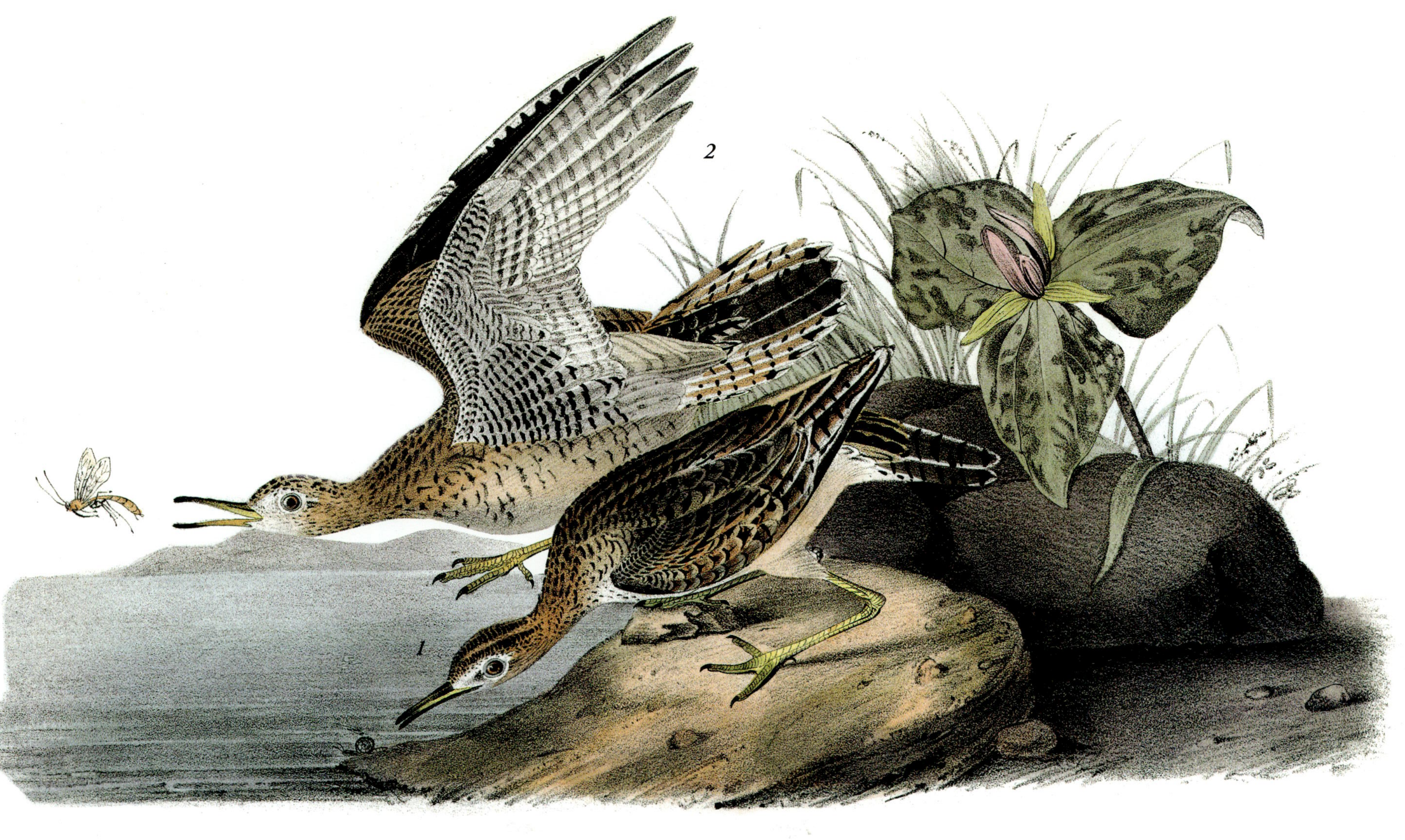

95. Upland Sandpiper

1. Male 2. Female

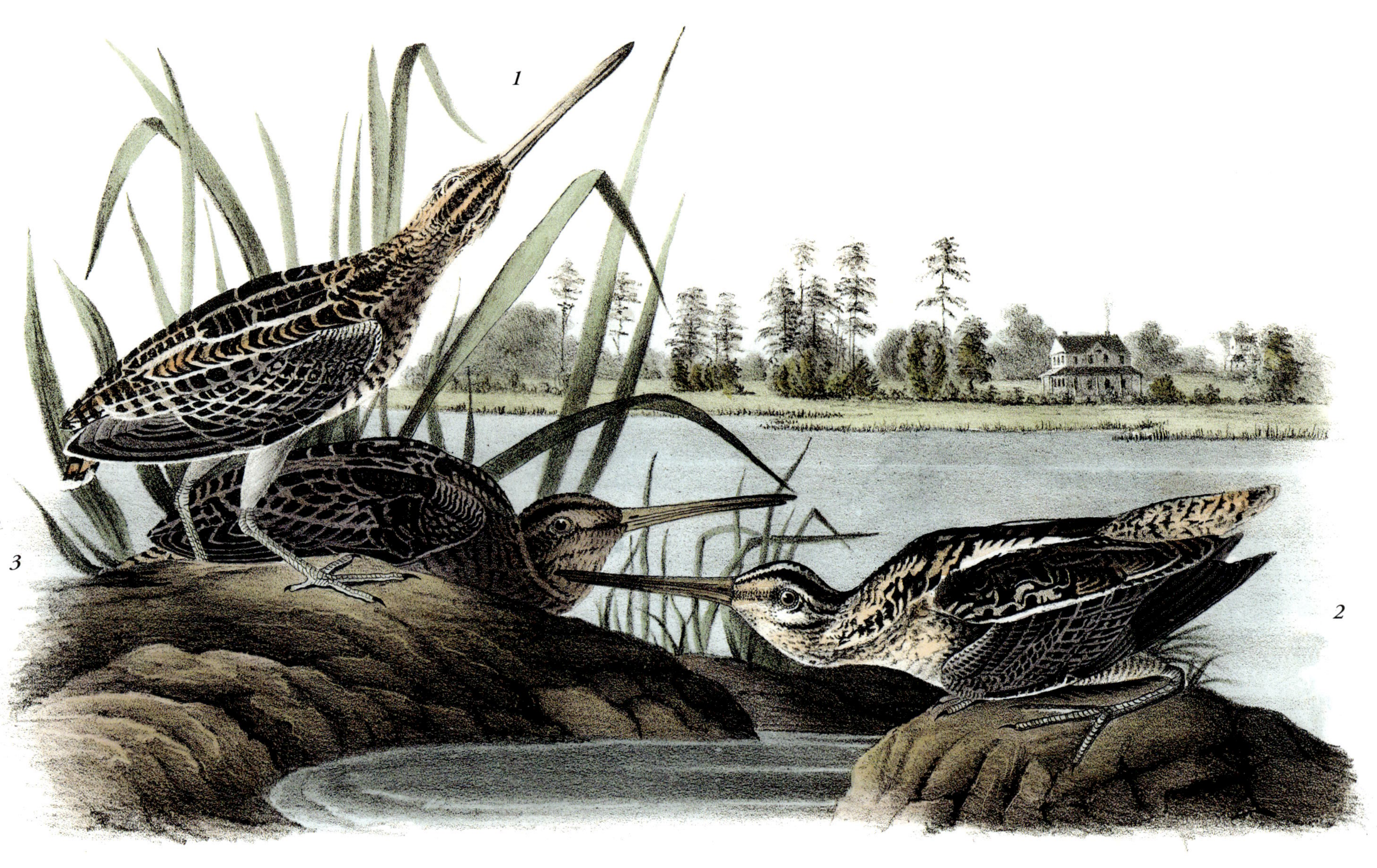

96. Common Snipe

1. Male   2, 3. Females

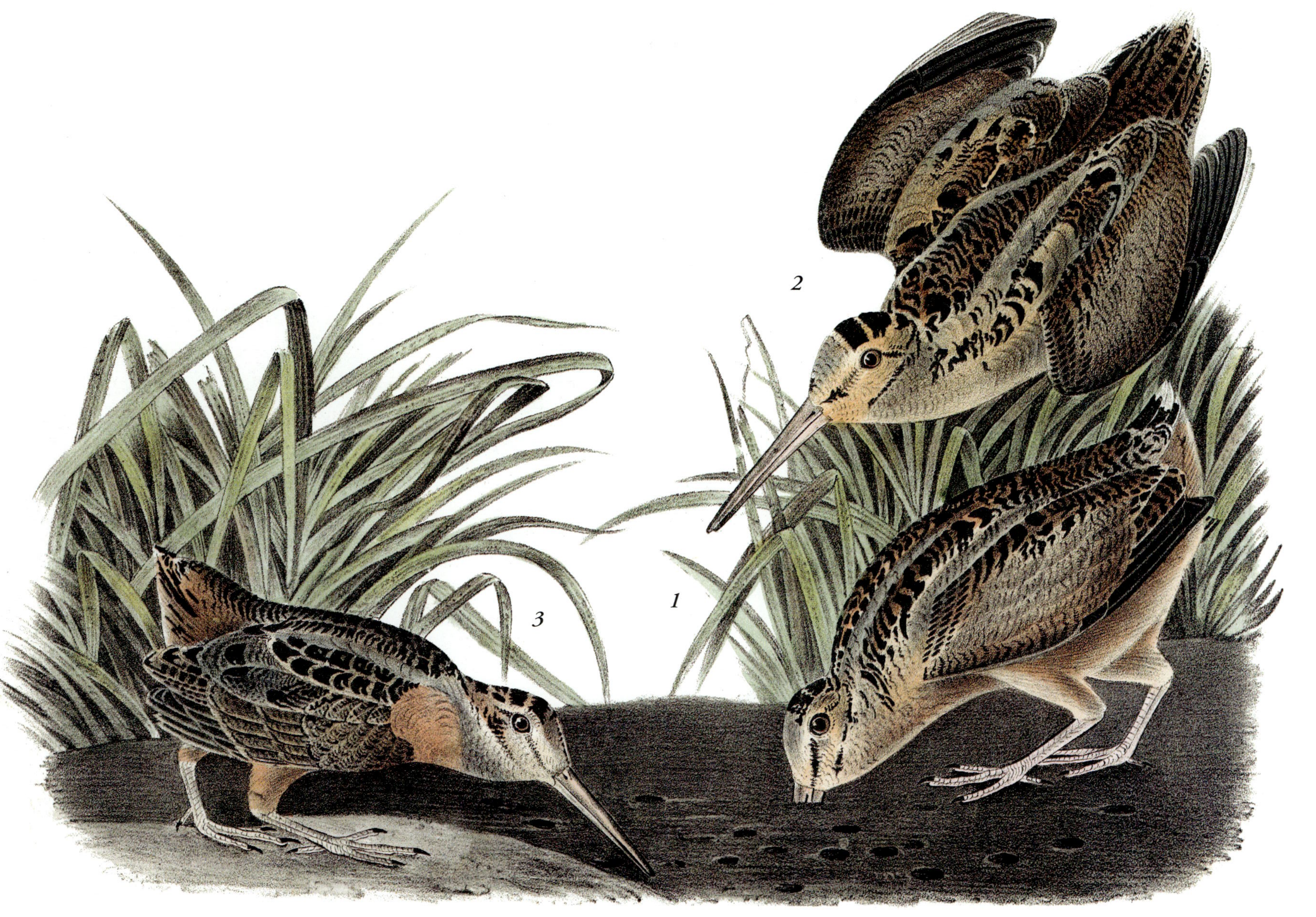

97. American Woodcock

1. Male 2. Female 3. Young in Autumn

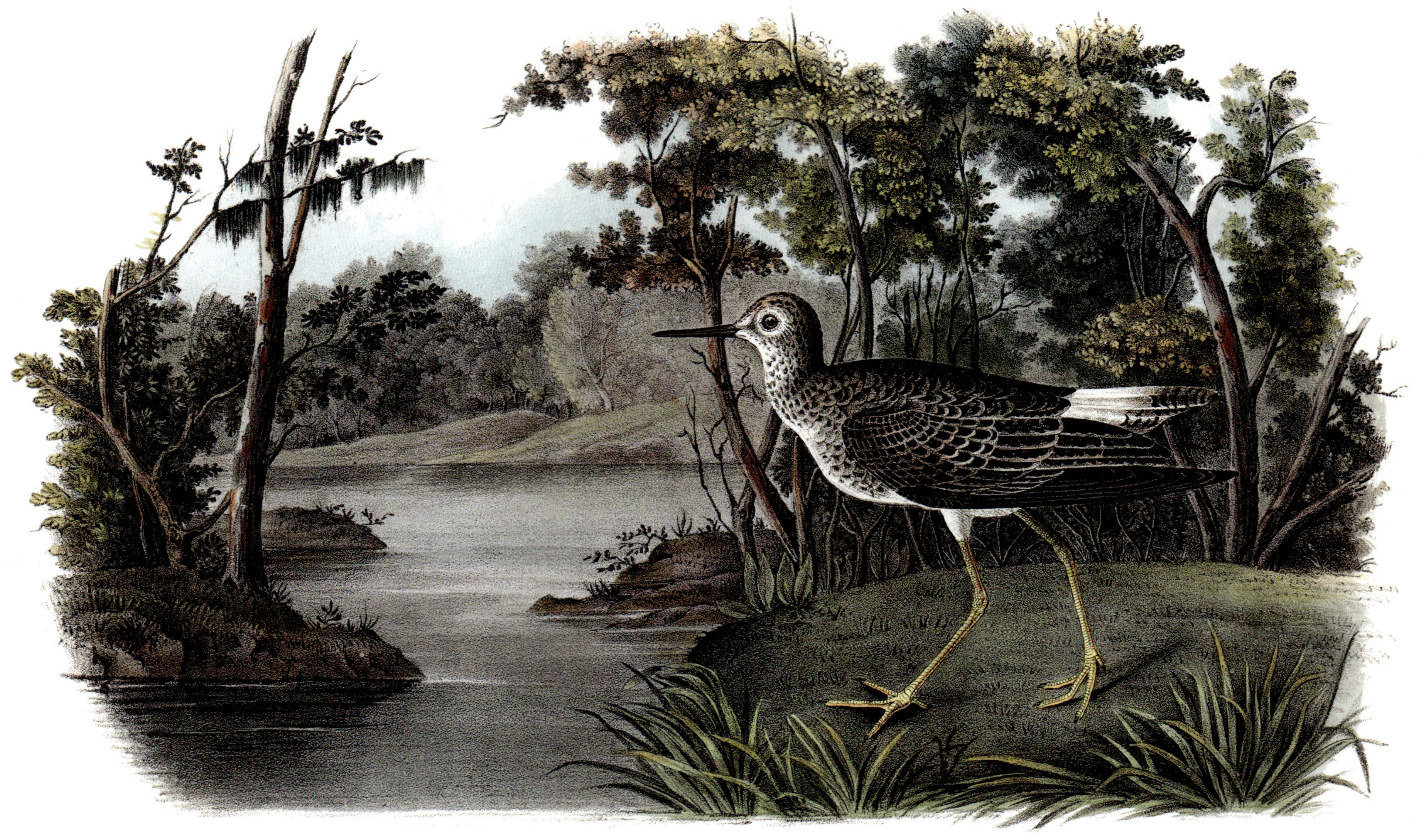

98. Lesser Yellowlegs
Male, Summer Plumage

99. Long-Billed Curlew

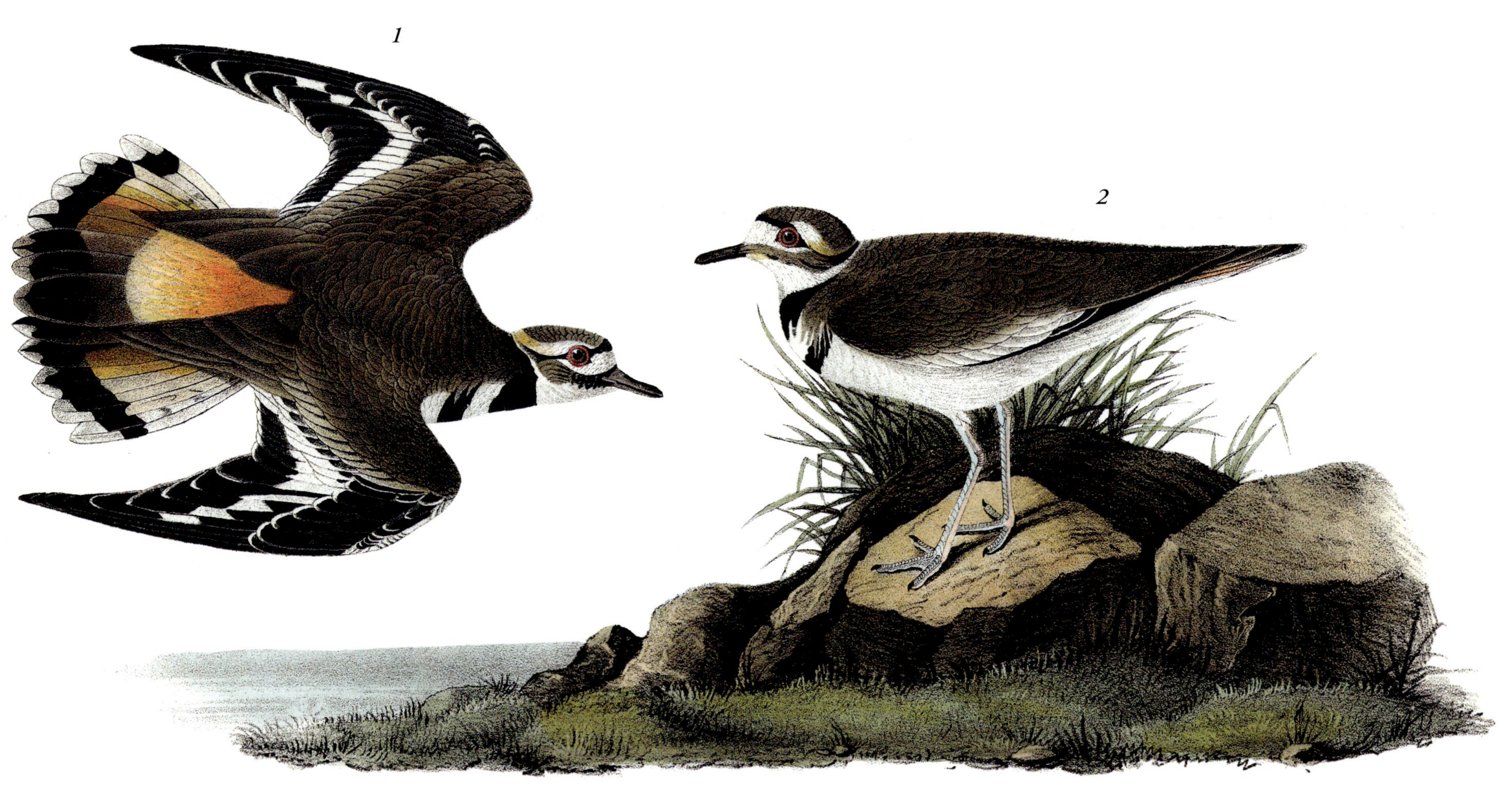

100. Killdeer

1. Male 2. Female

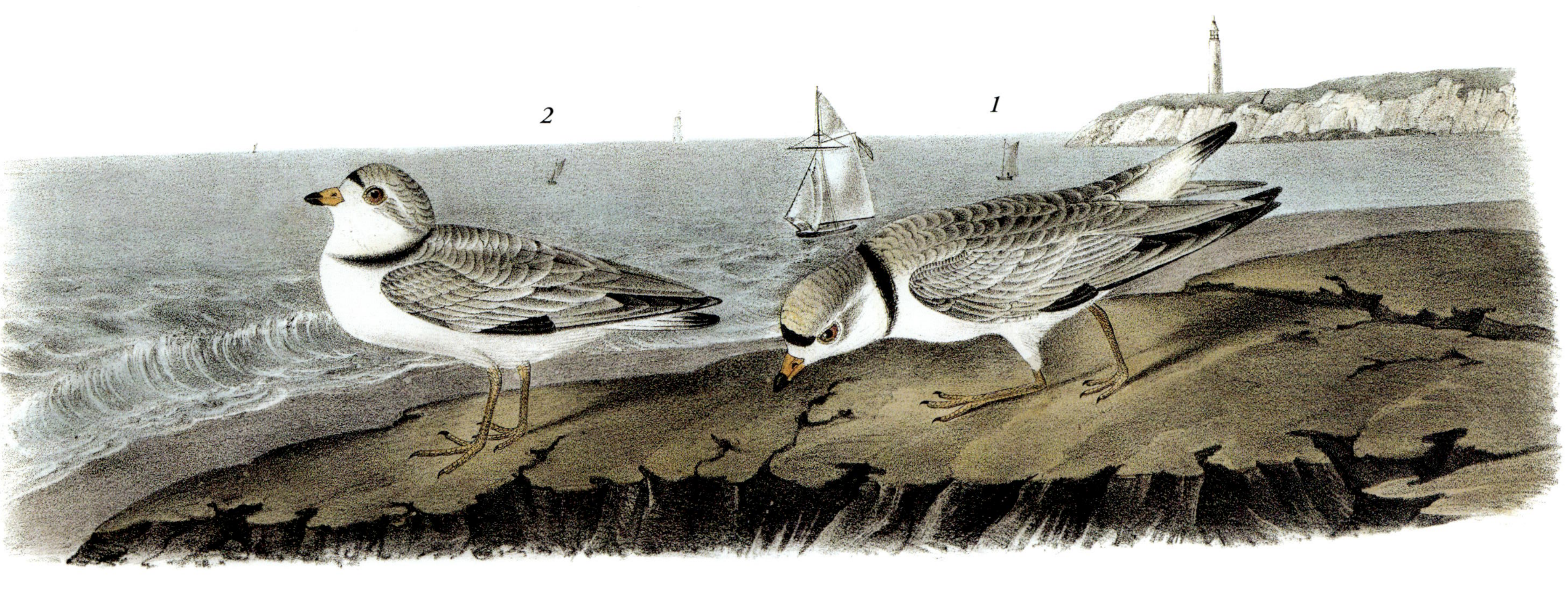

101. Piping Plover

1. Male 2. Female

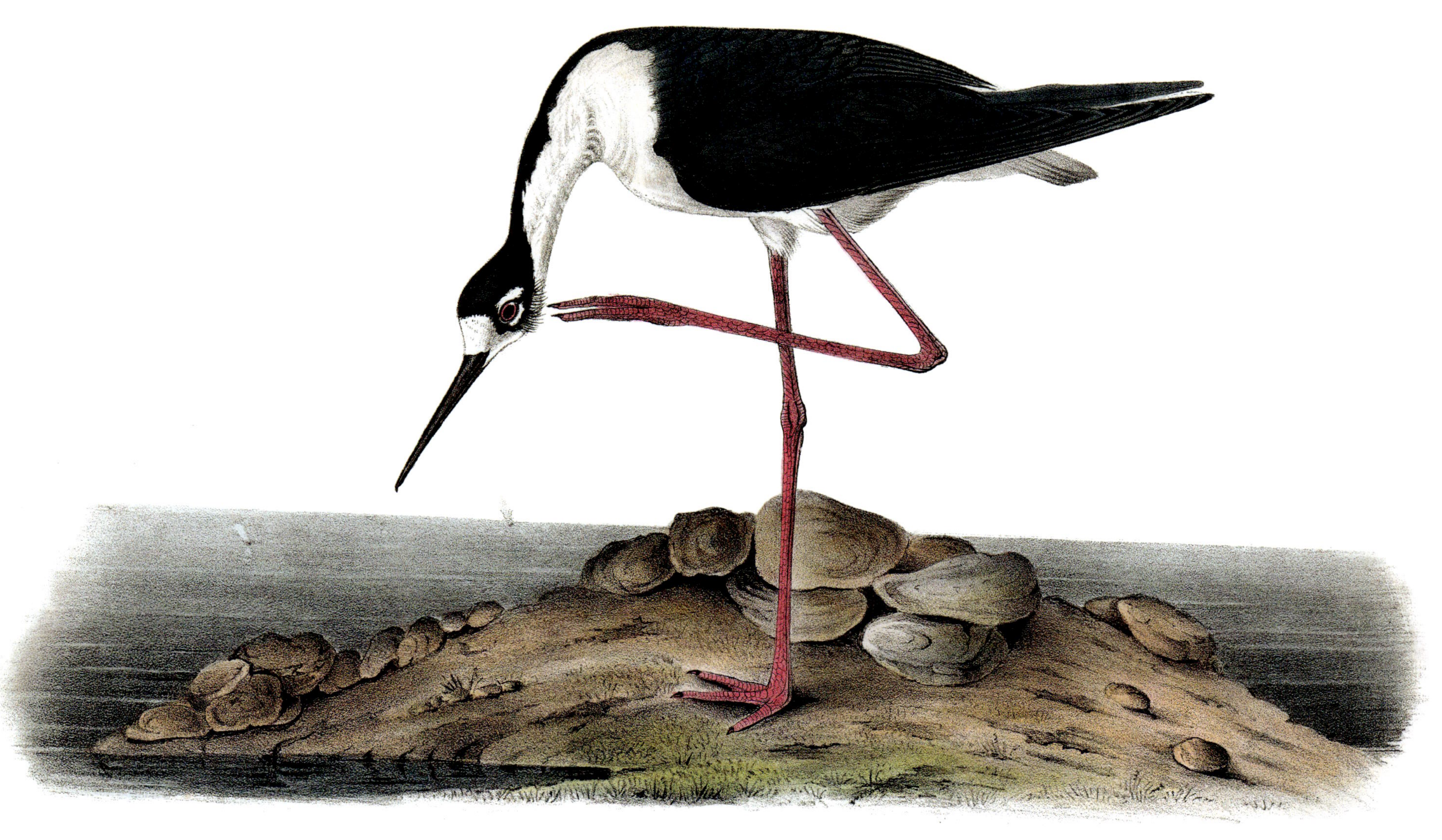

102. Black-Necked Stilt

Male

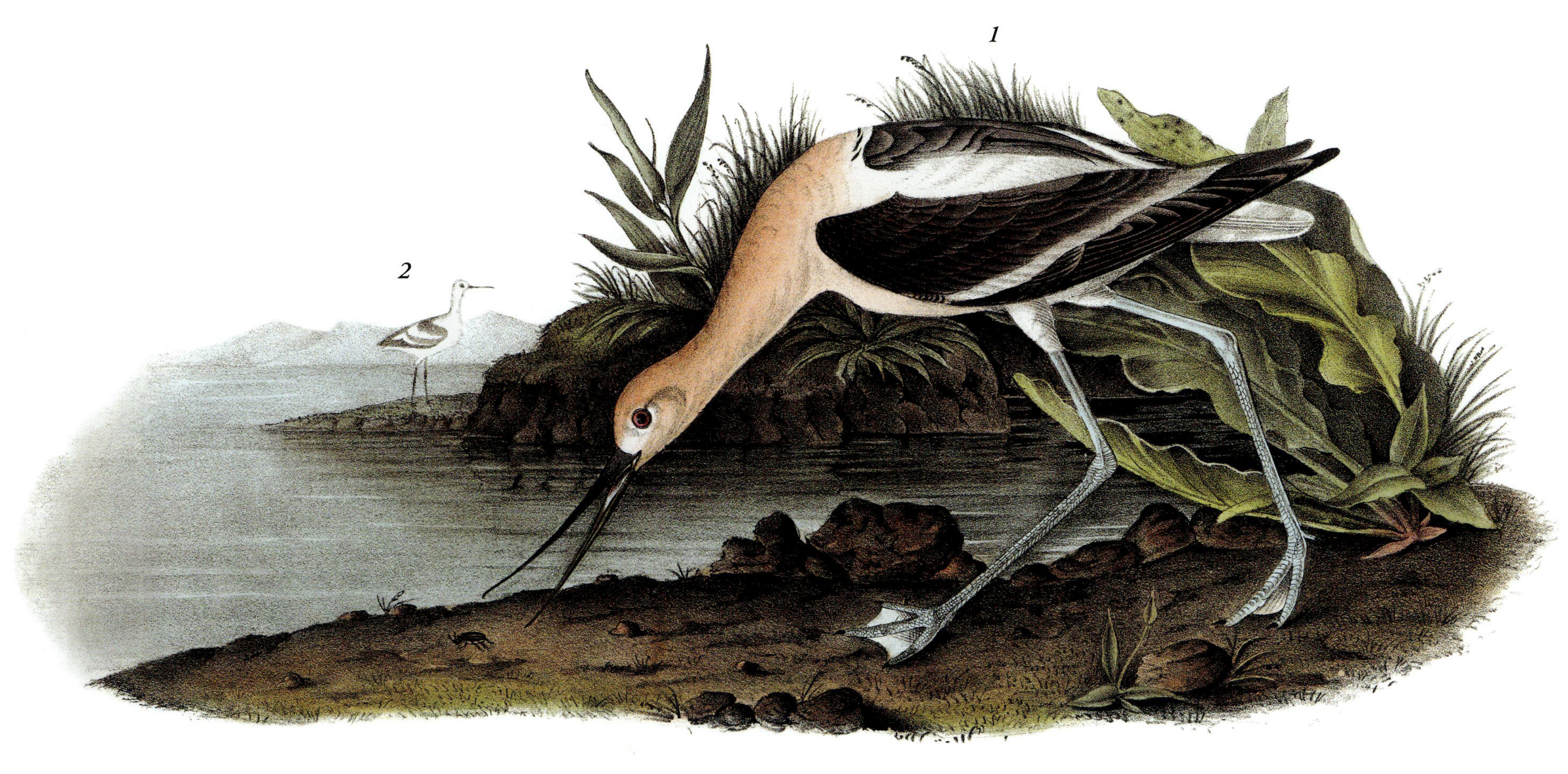

103. American Avocet

1. Young, First Winter Plumage 2. Adult

104. American Oystercatcher

Male

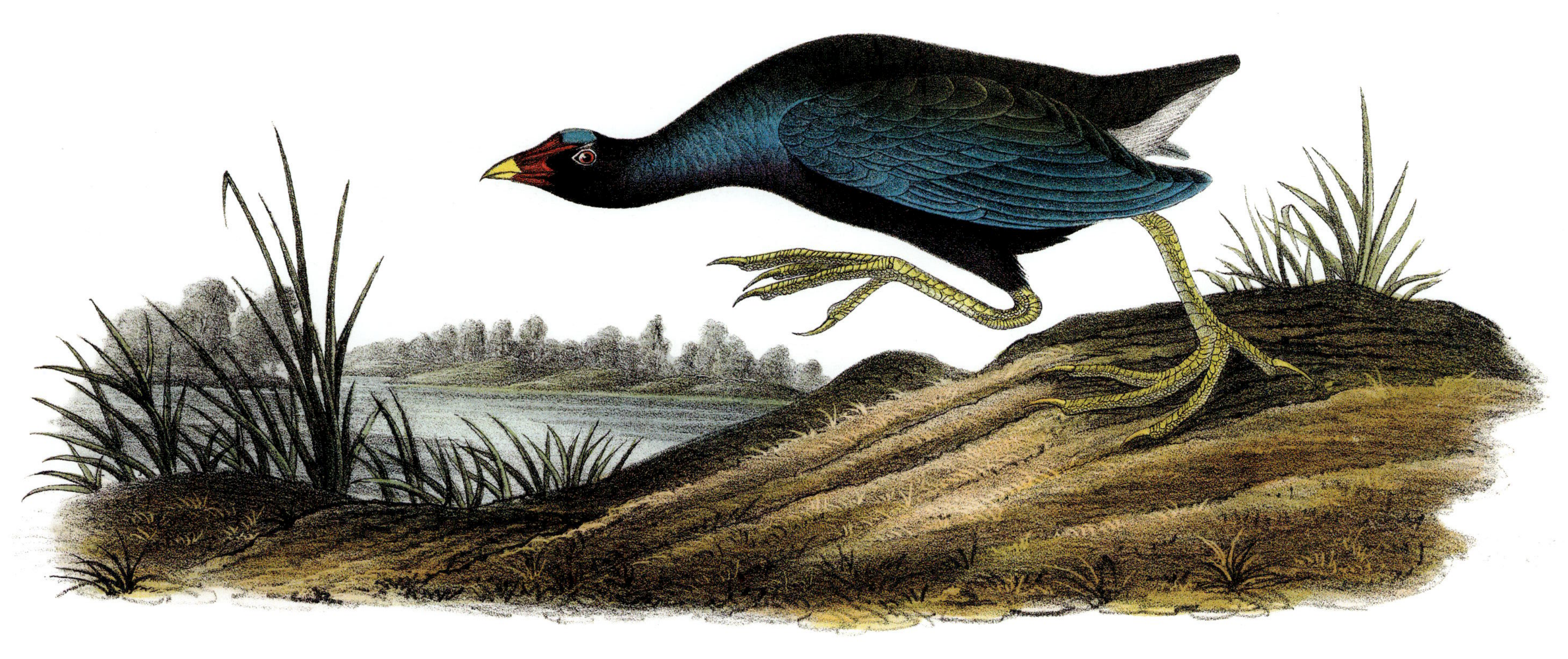

105. Purple Gallinule

Adult Male, Spring Plumage

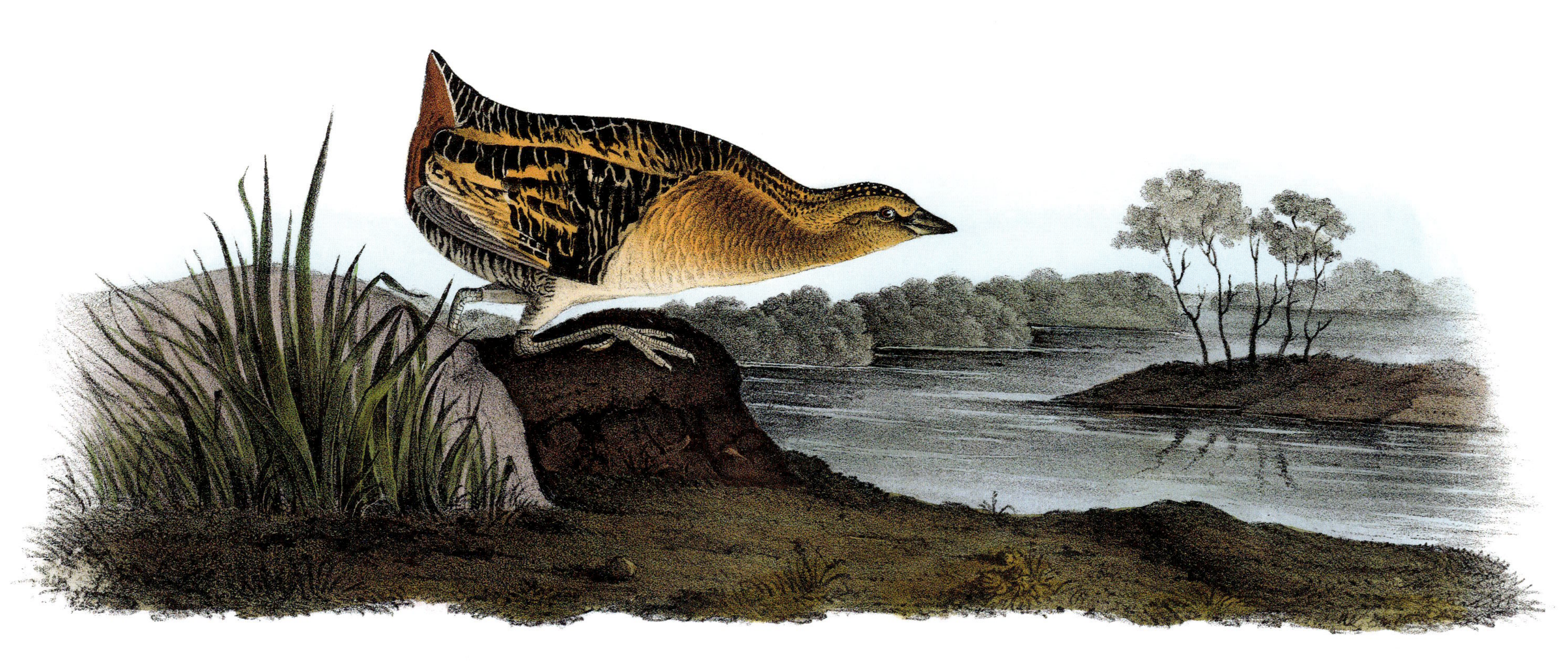

106. Yellow Rail

Adult Male in Spring

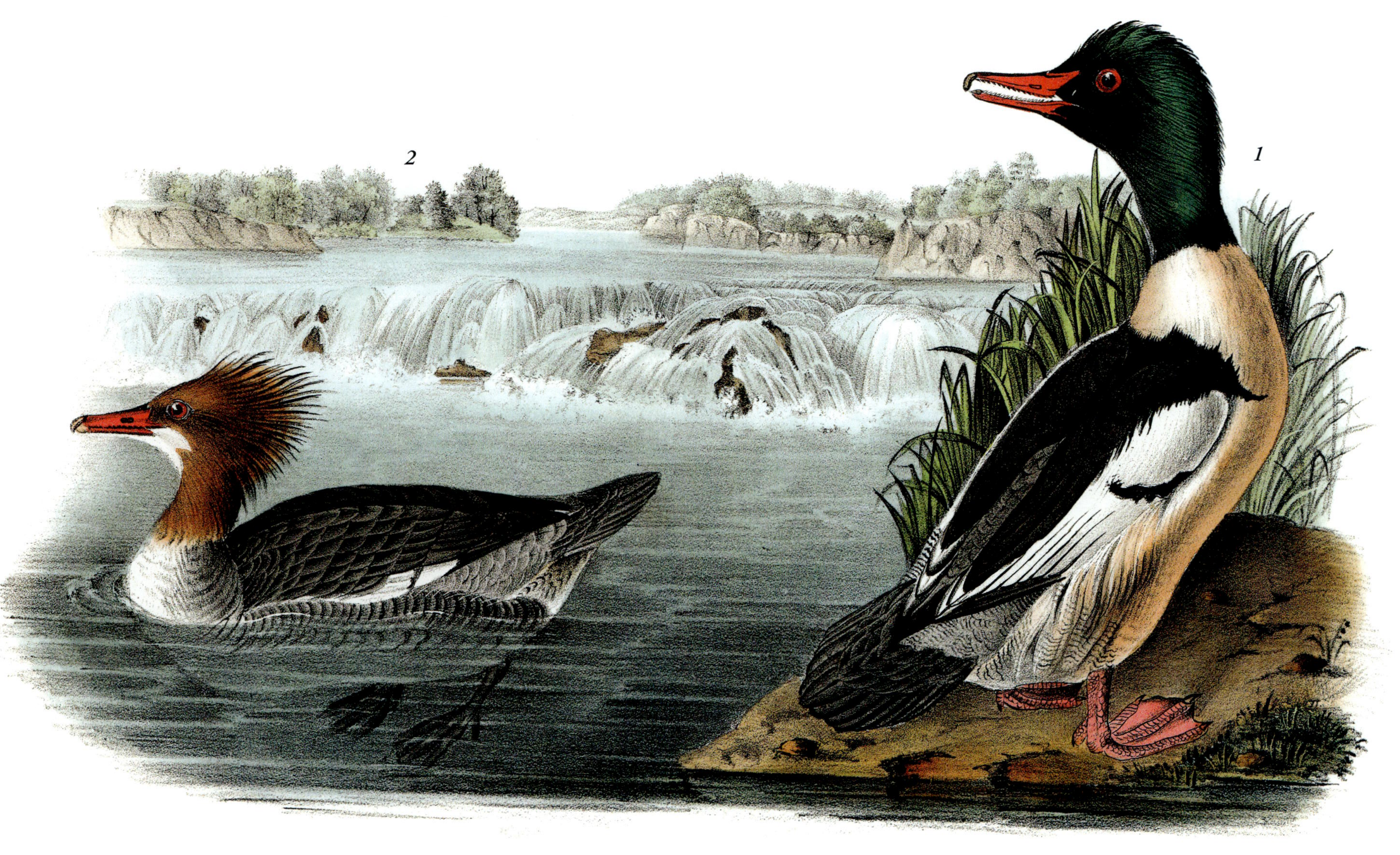

107. Common Merganser

1. Male 2. Female

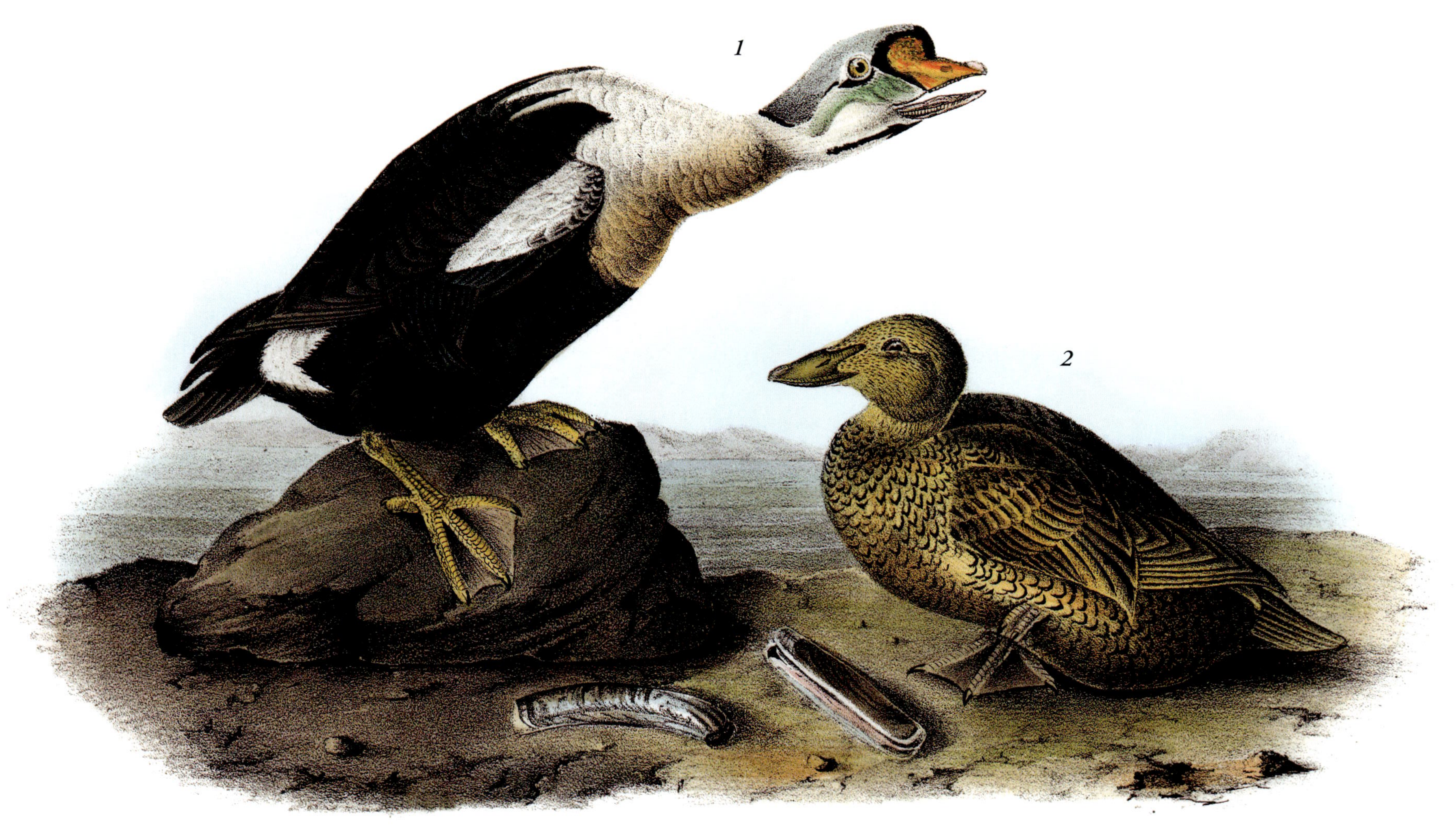

108. King Eider

1. Male 2. Female

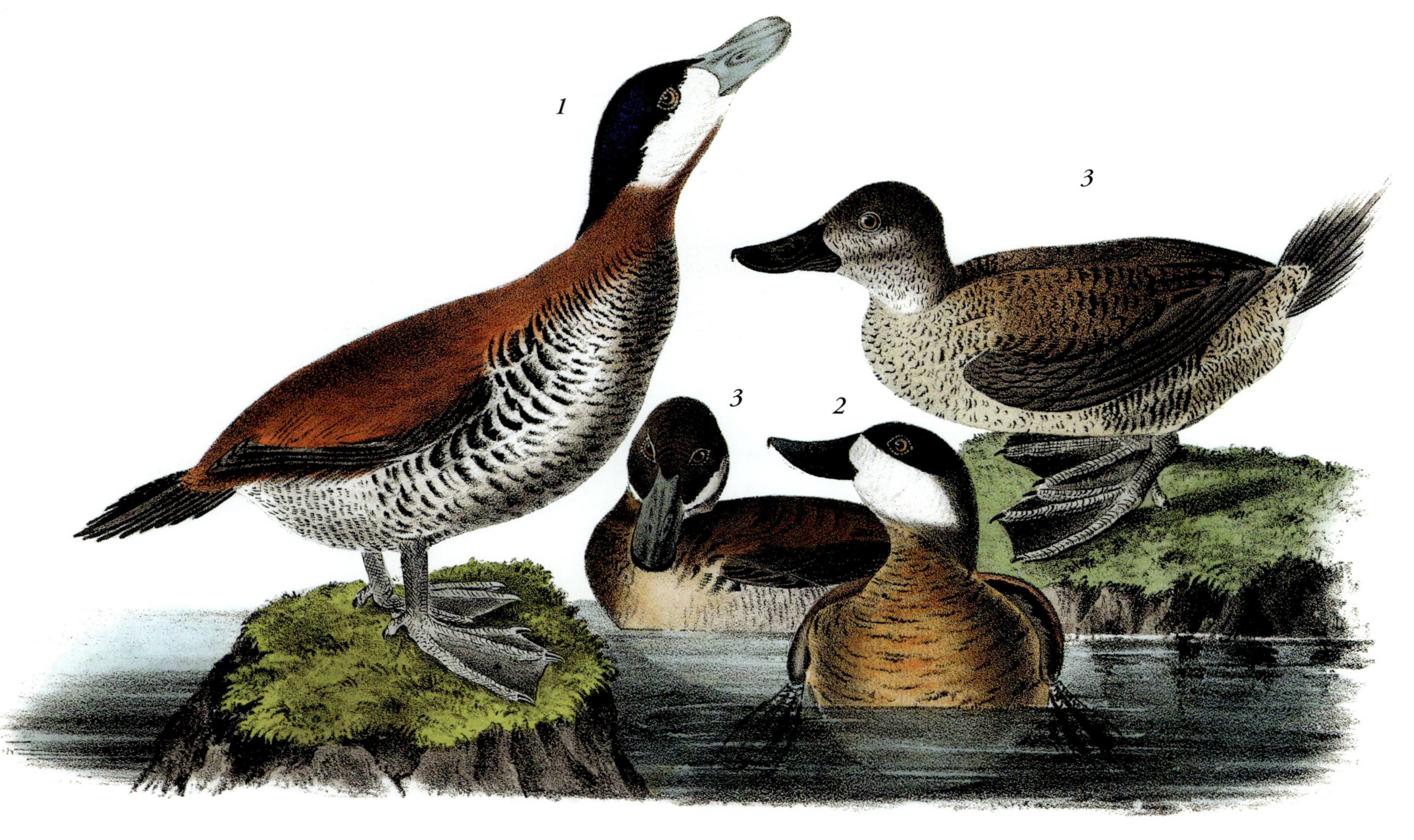

109. Ruddy Duck

1. Male 2. Female 3. Young

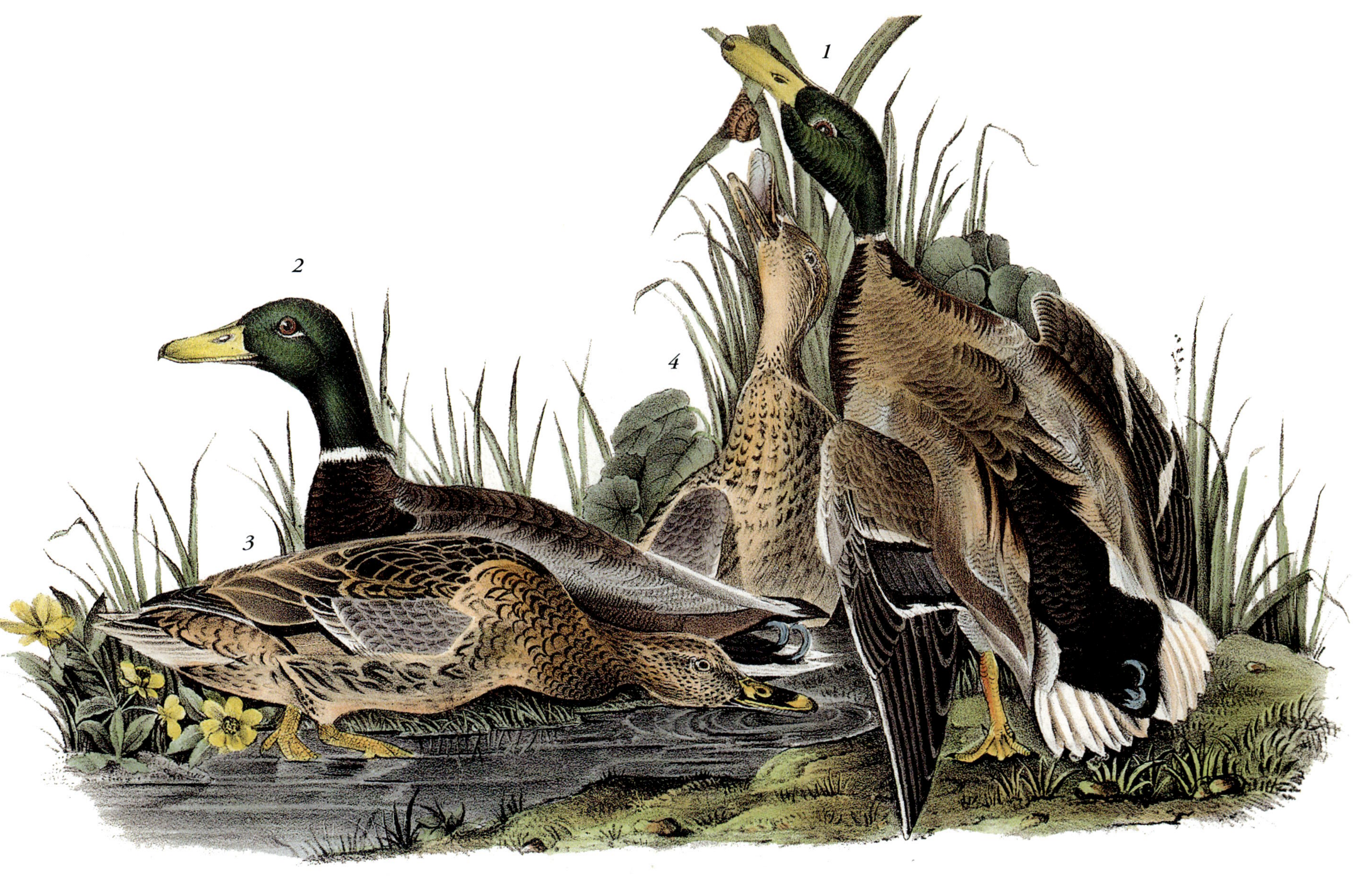

110. Mallard

1, 2. Males 3, 4. Females

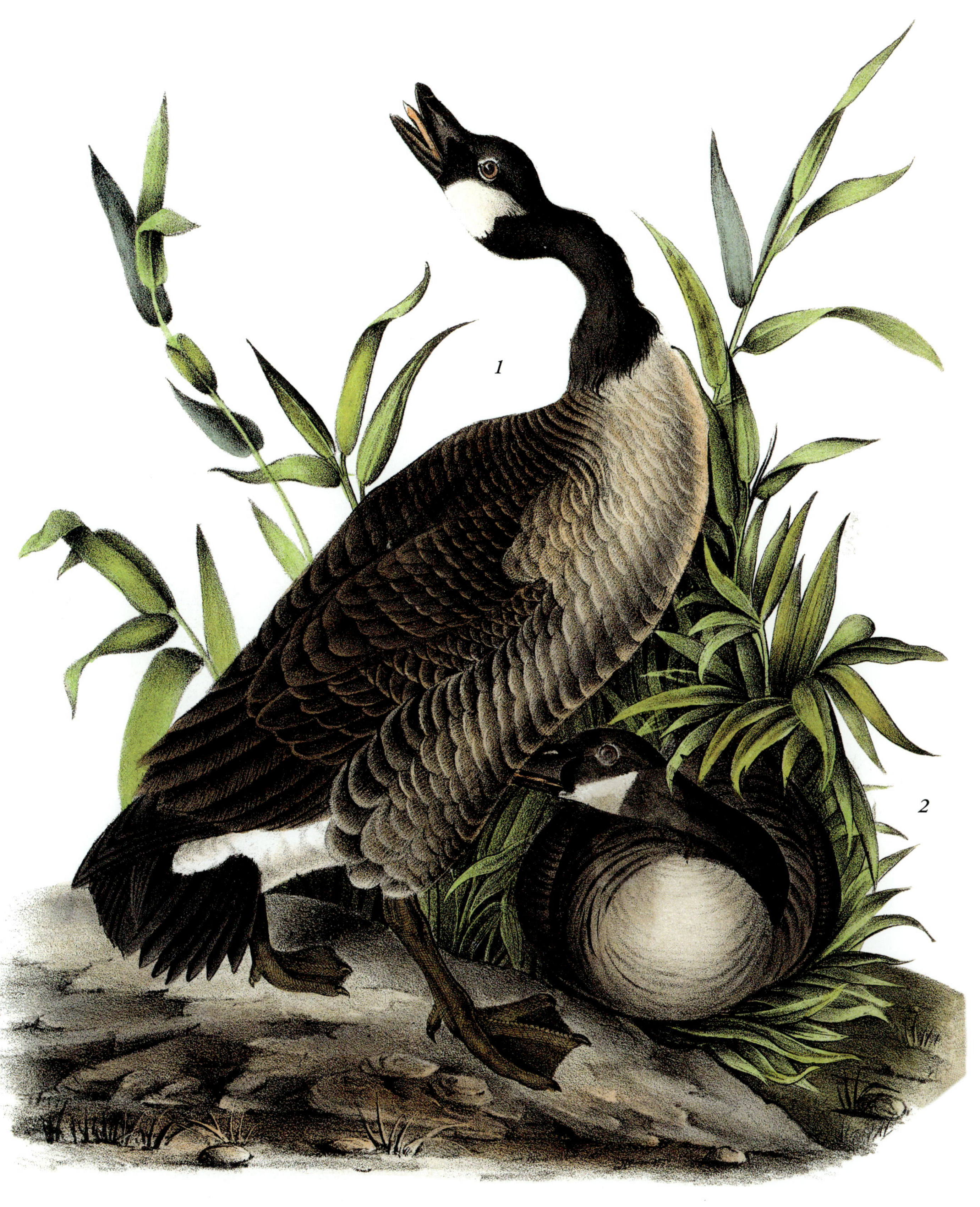

111. Canada Goose

1. Male 2. Female

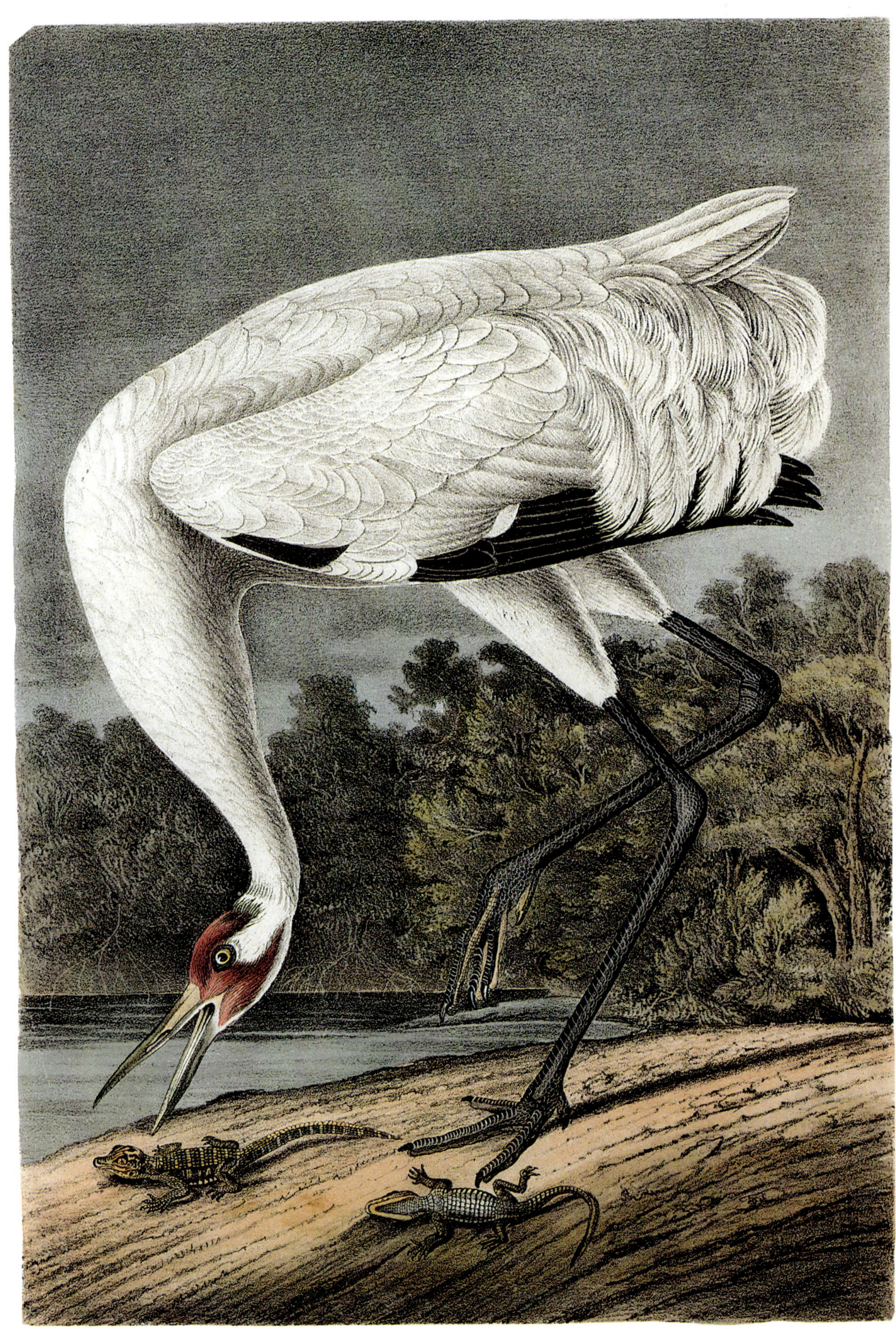

112. Whooping Crane

Adult Male

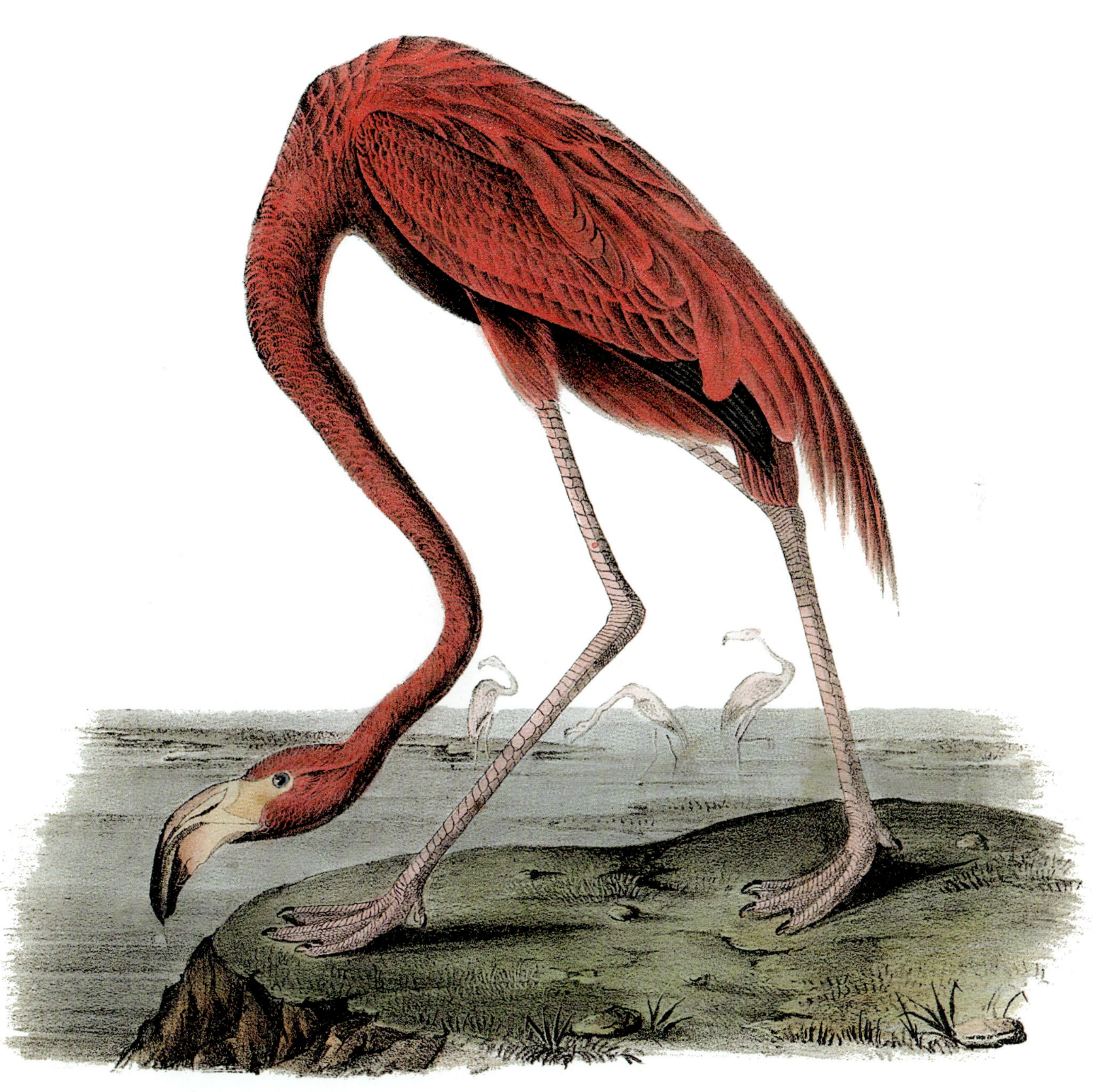

113. Greater Flamingo

Adult Male

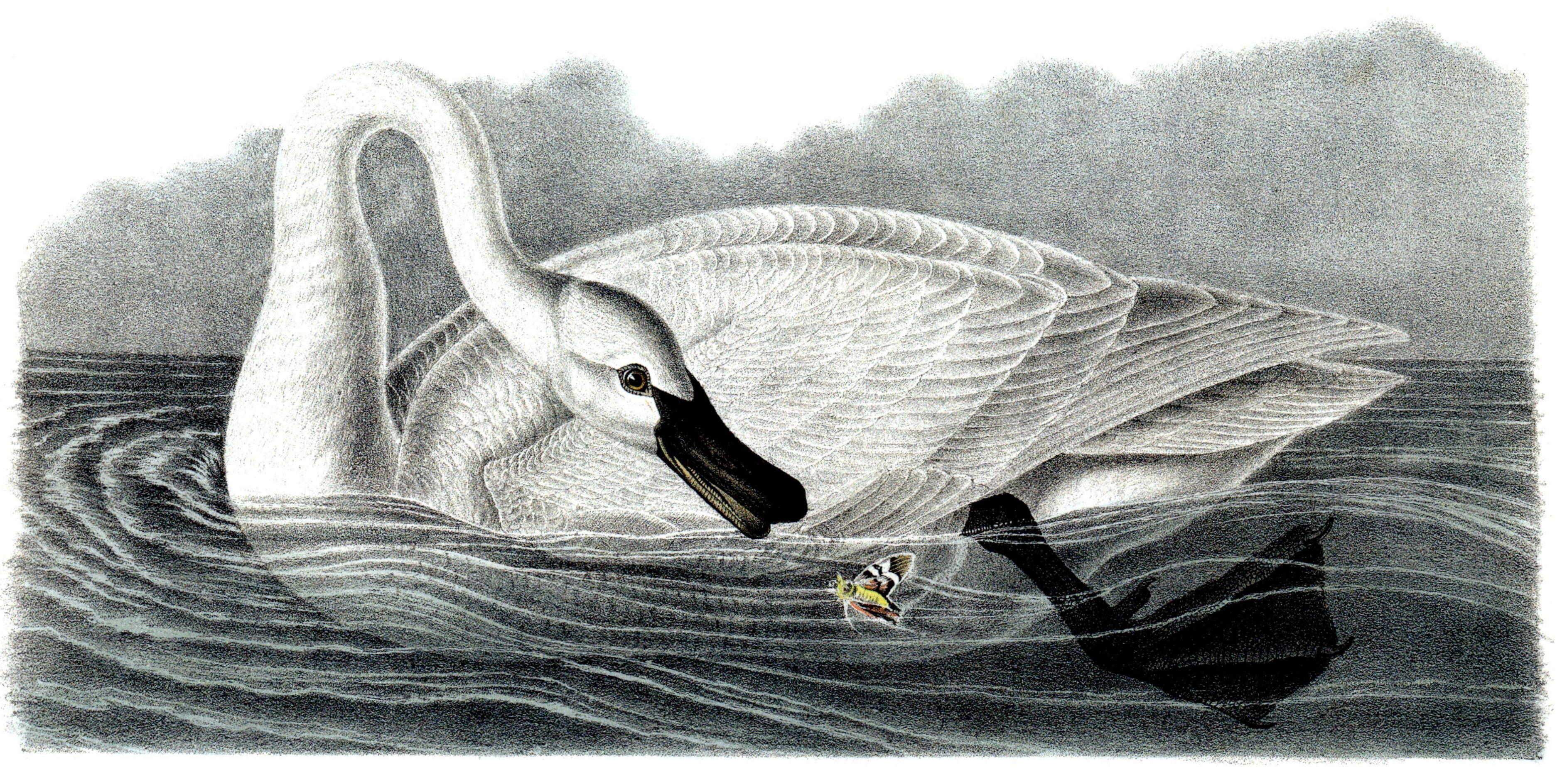

114. Trumpeter Swan

Adult

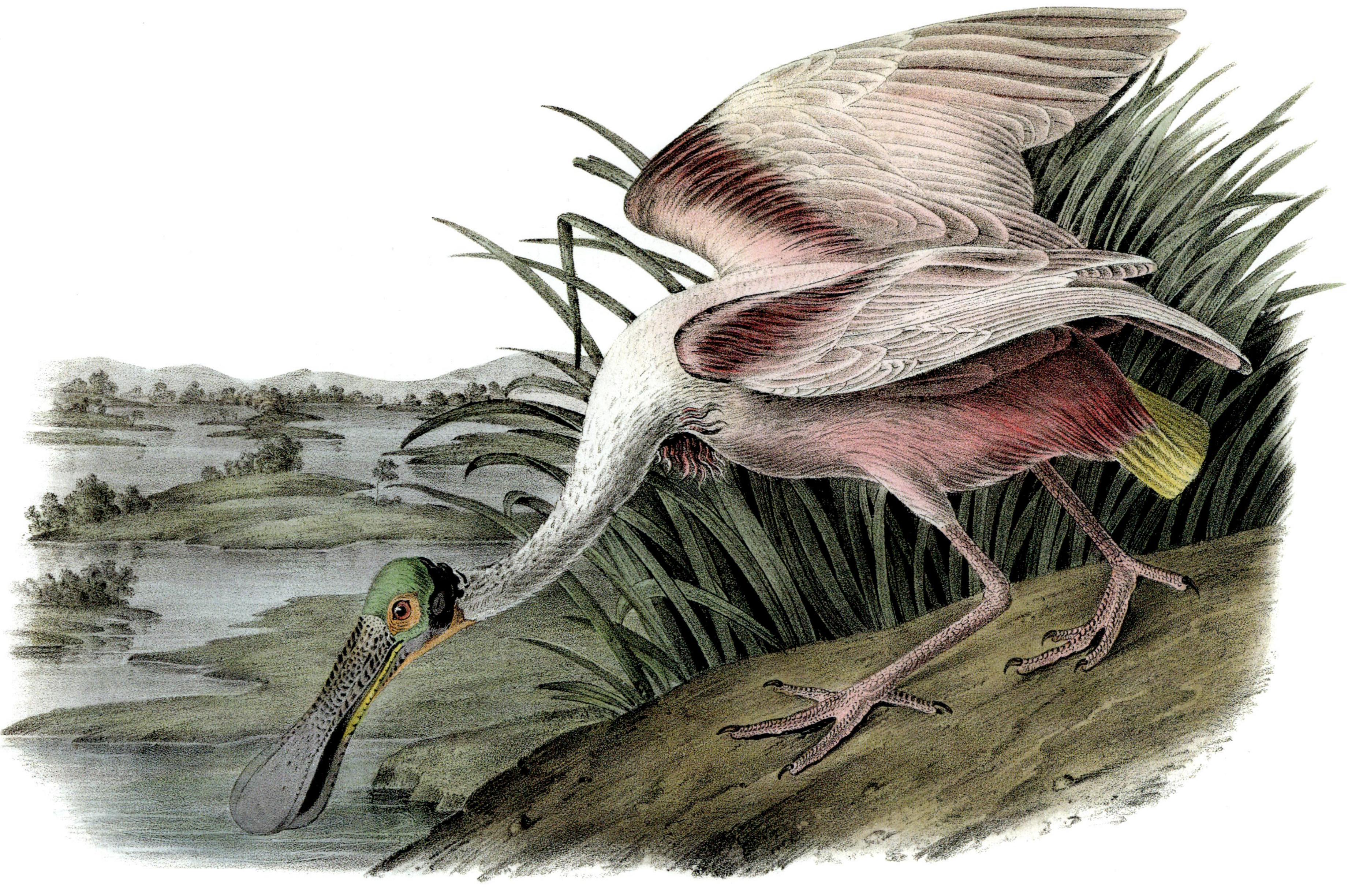

115. Roseate Spoonbill

Male

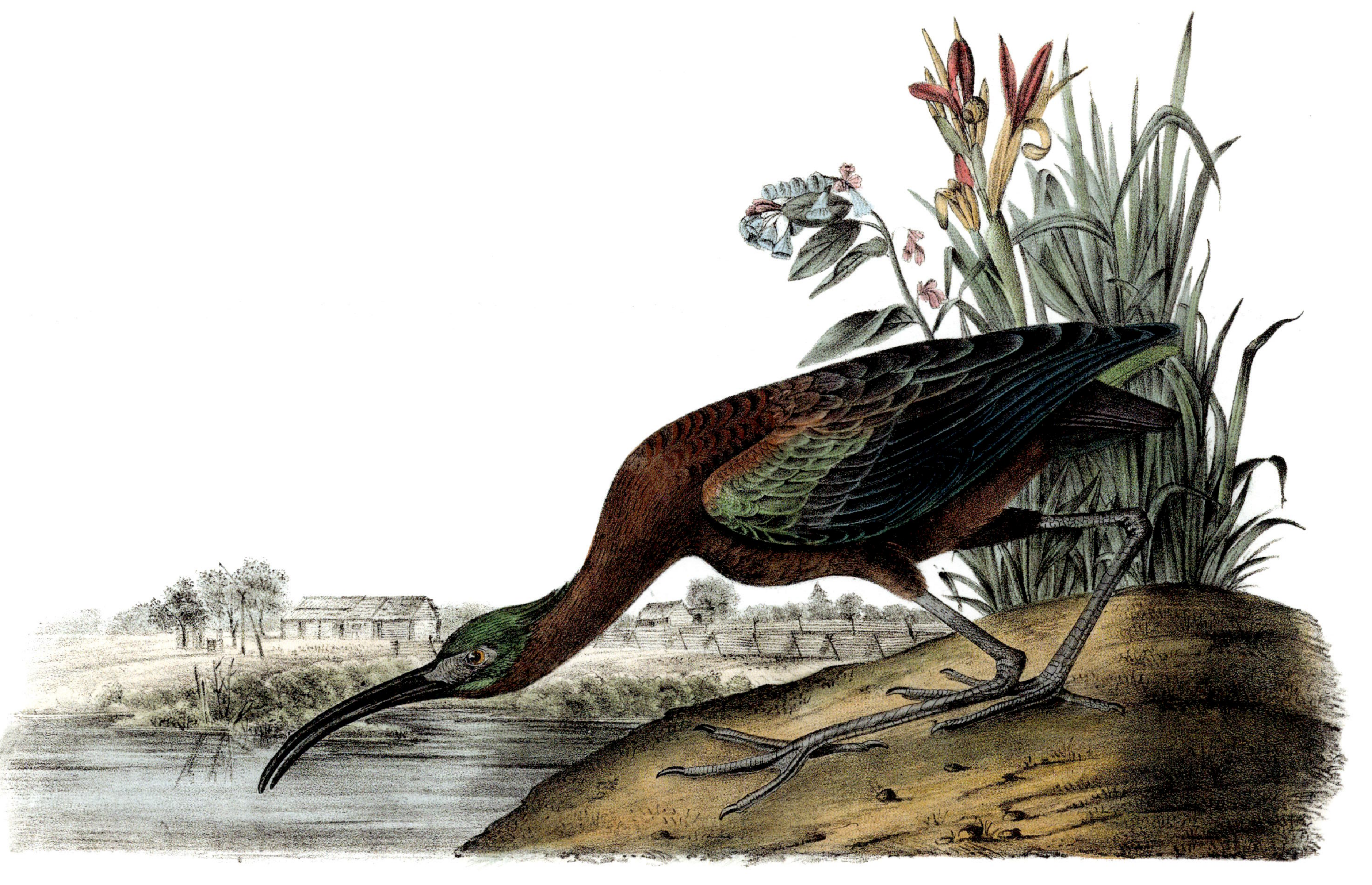

116. Glossy Ibis

Adult Male

117. Limpkin

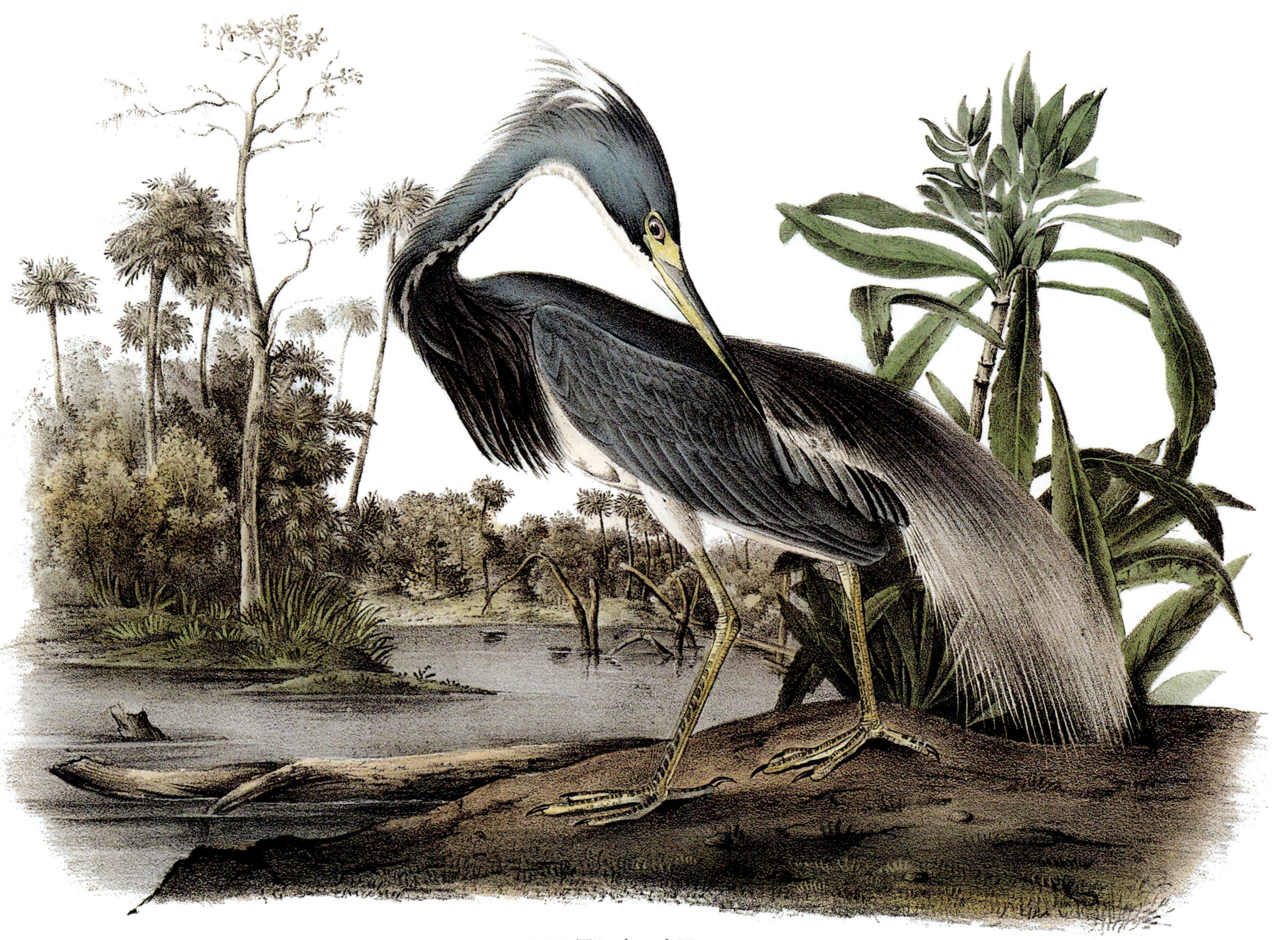

118. Tricolored Heron

Adult Male

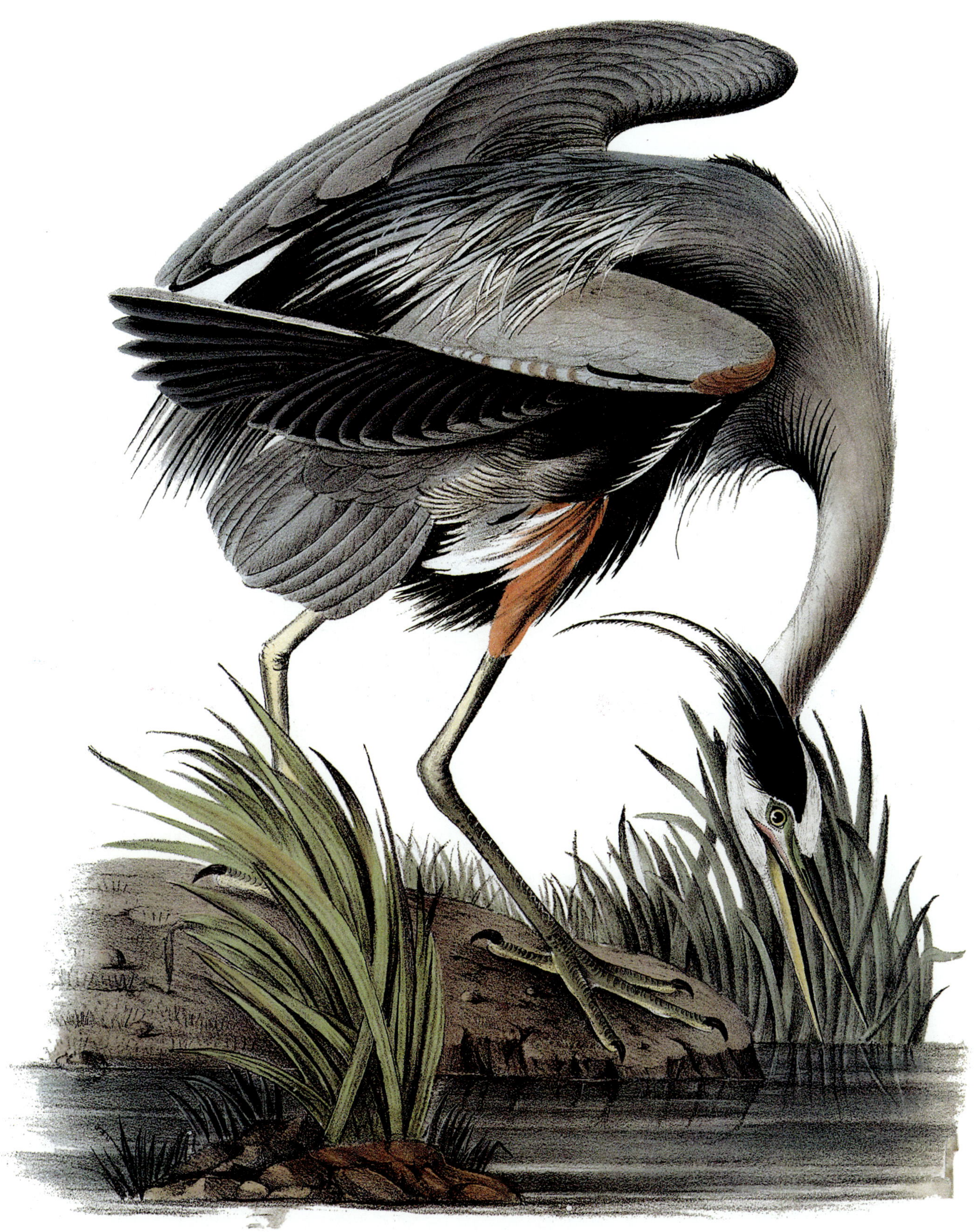

119. Great Blue Heron

Male

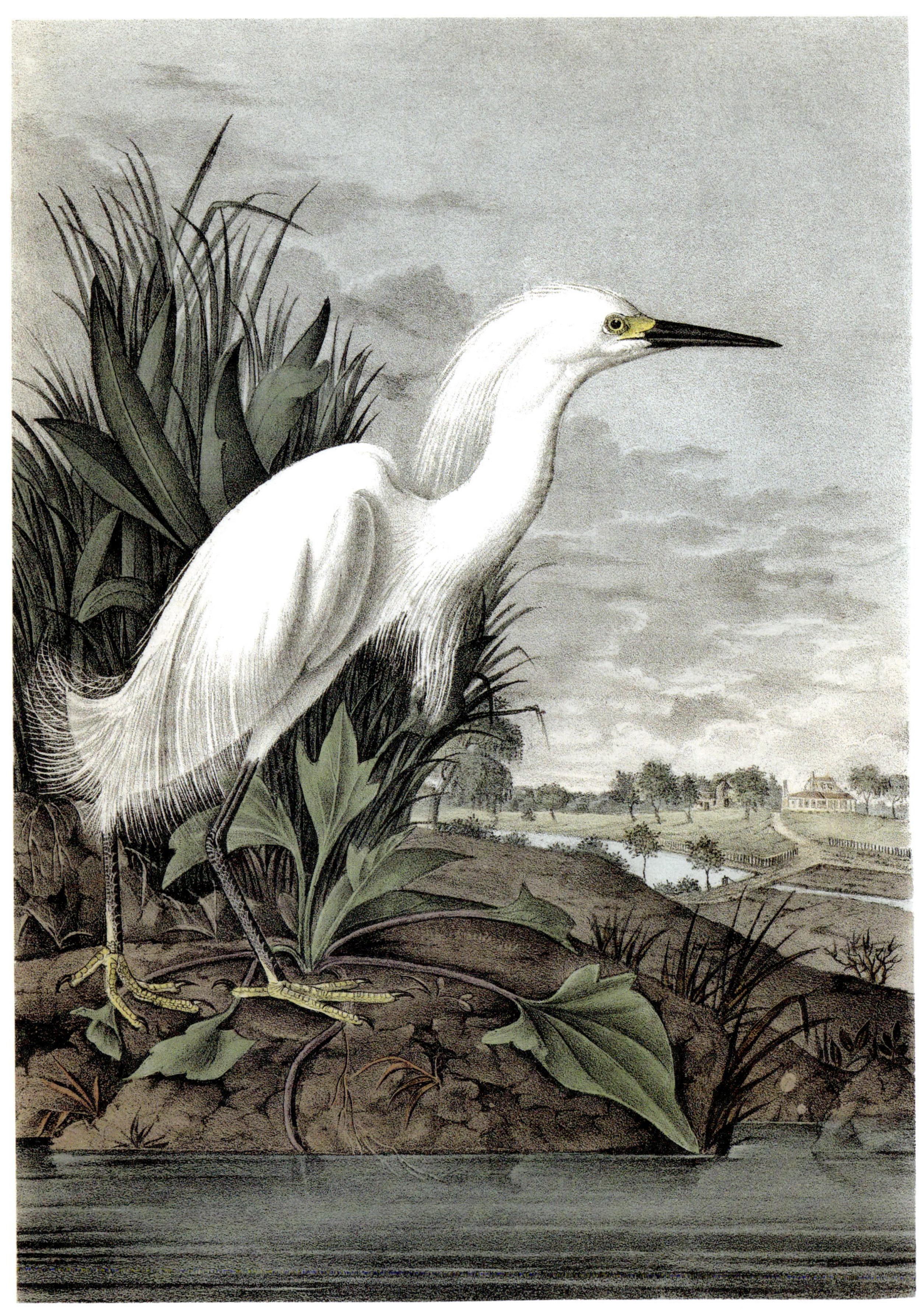

120. Snowy Egret

Male

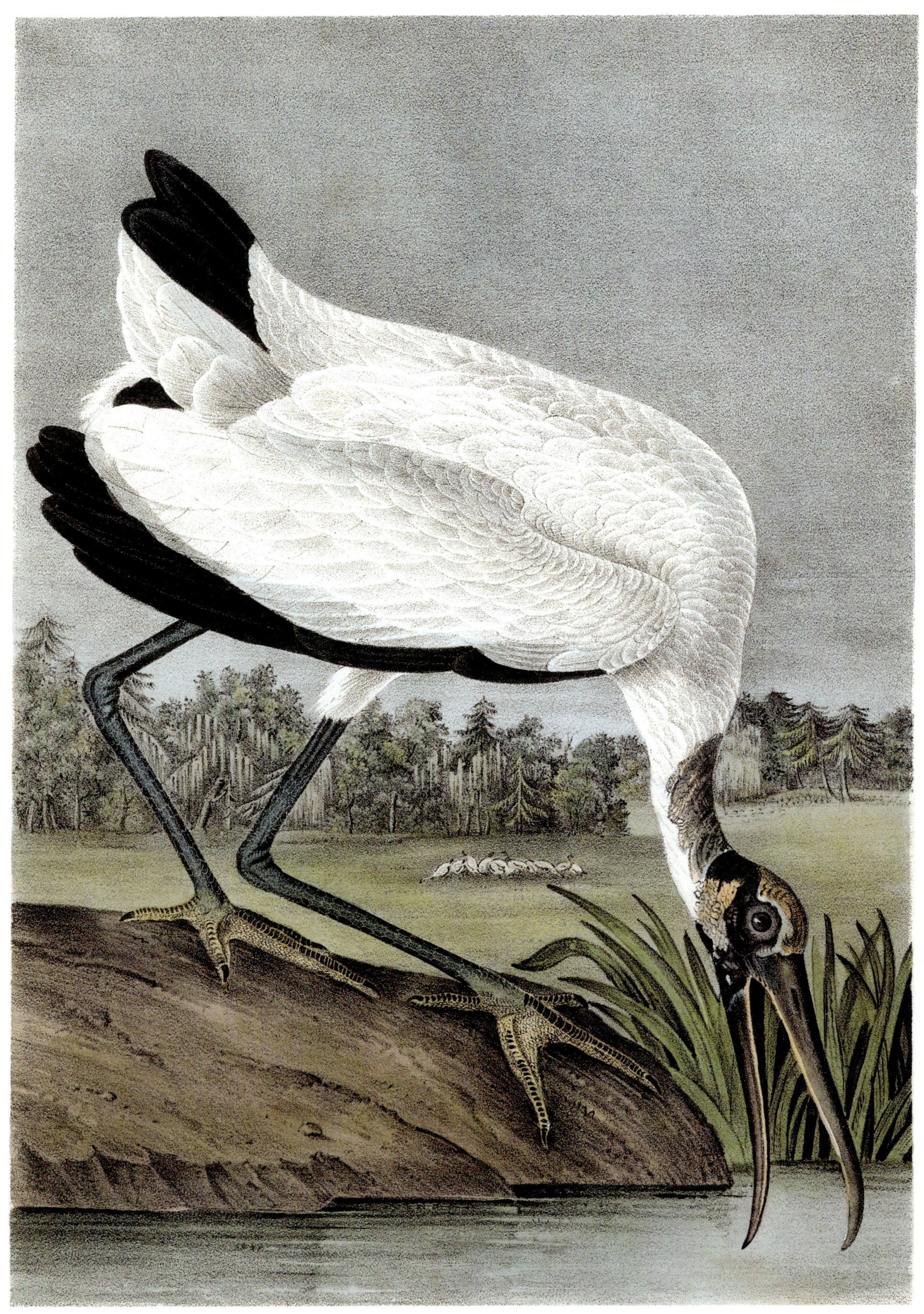

121. Wood Stork

Male

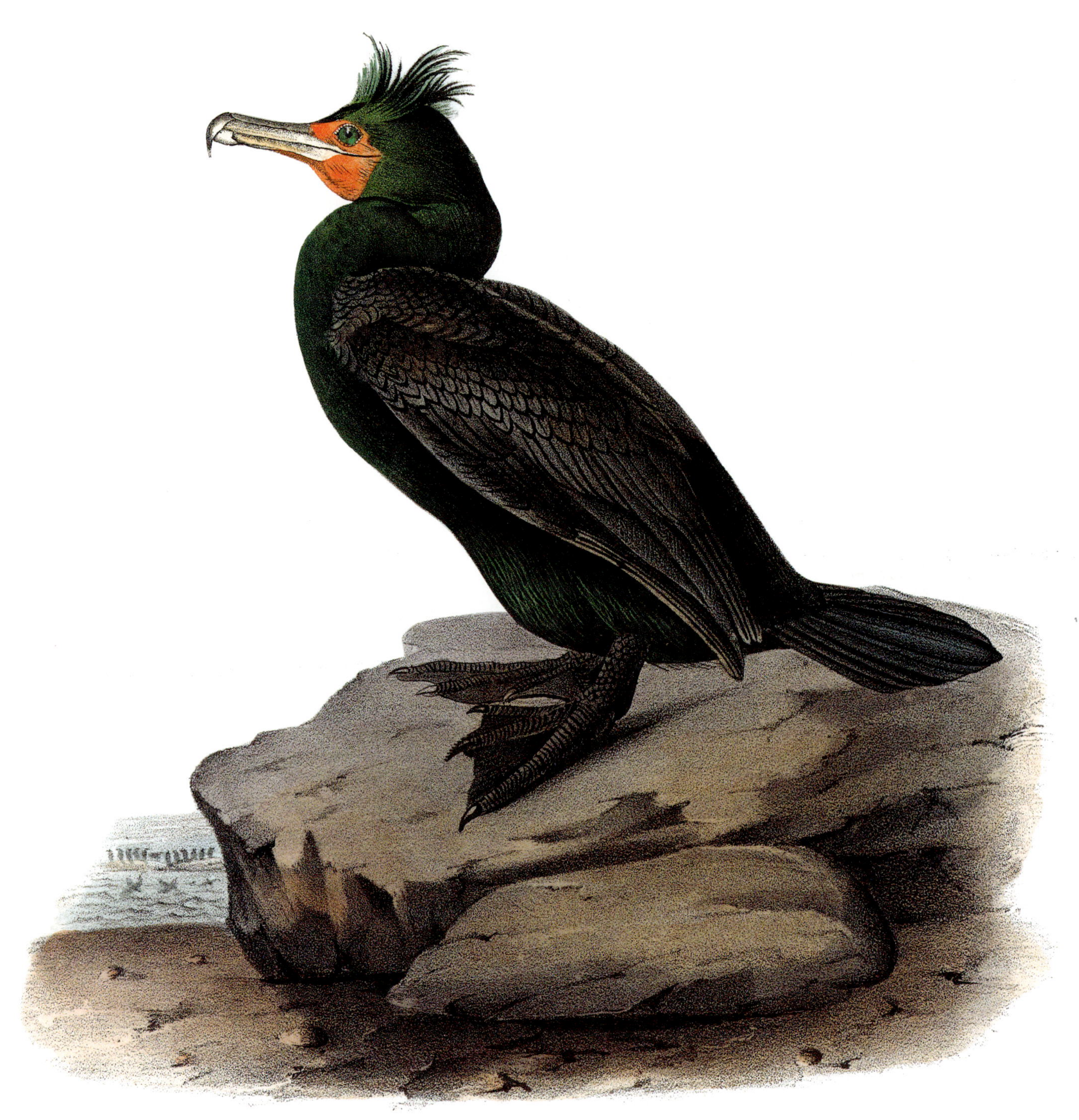

122. Double-Crested Cormorant

Male

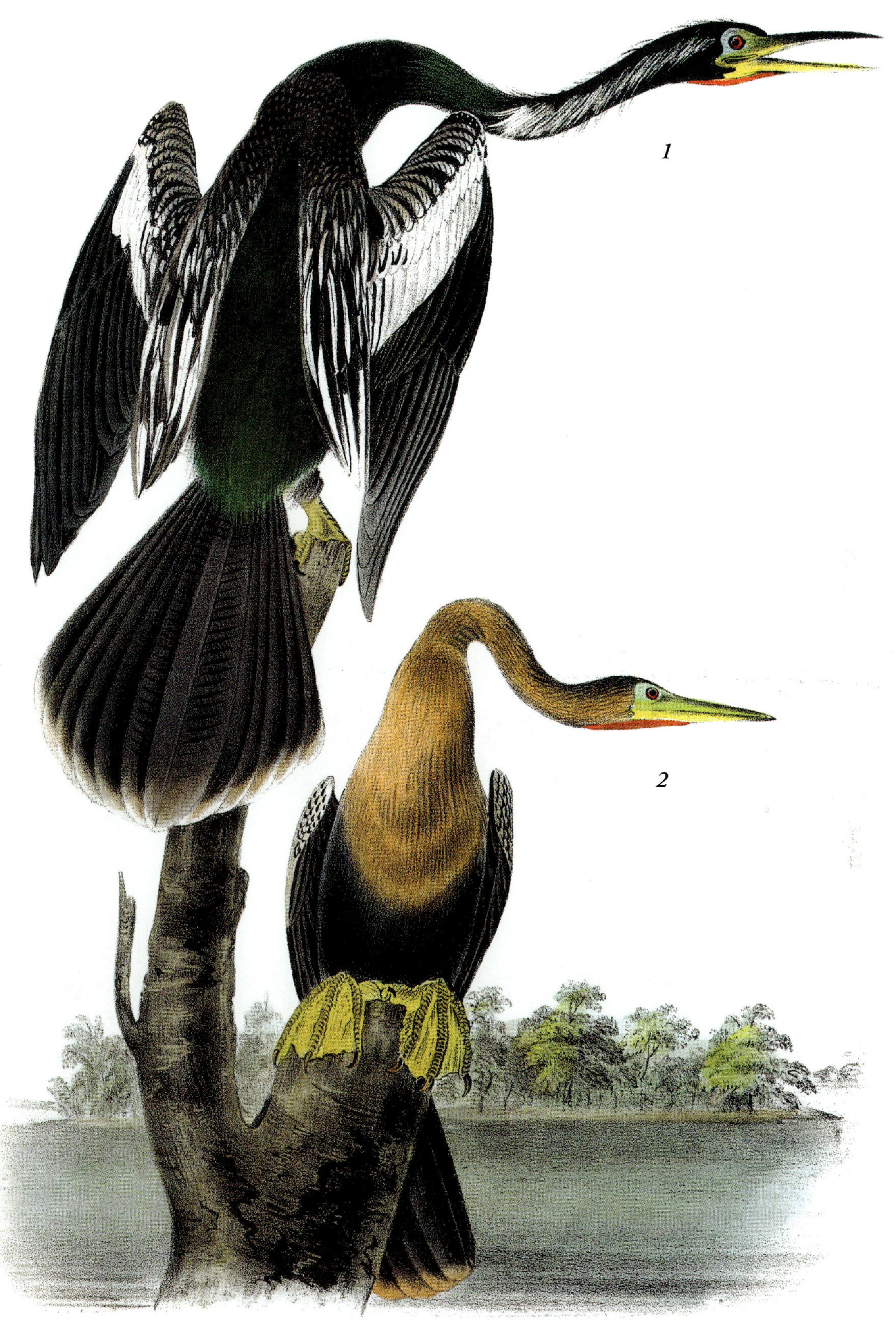

123. Anhinga

1. Male 2. Female

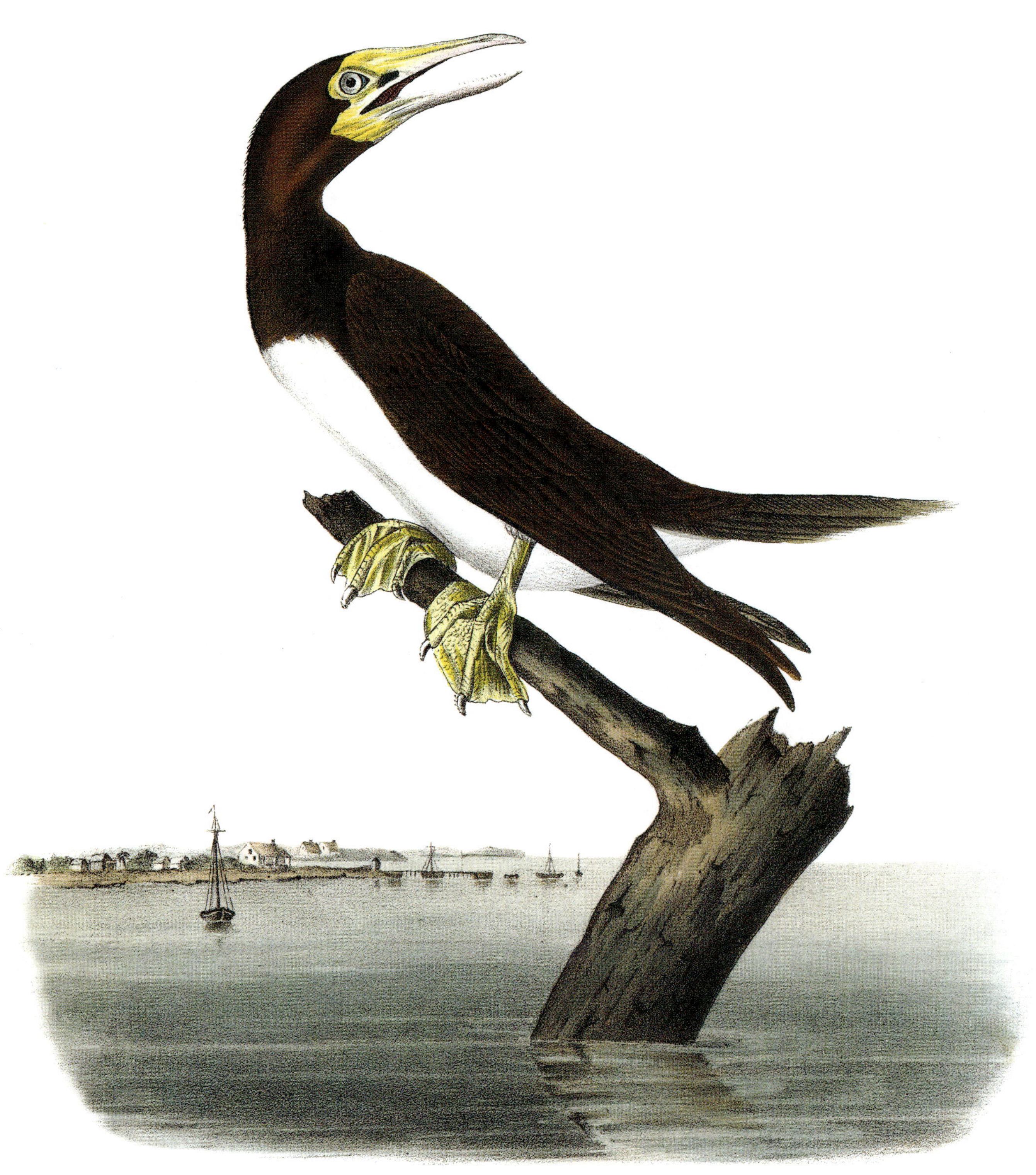

124. Brown Booby

Male

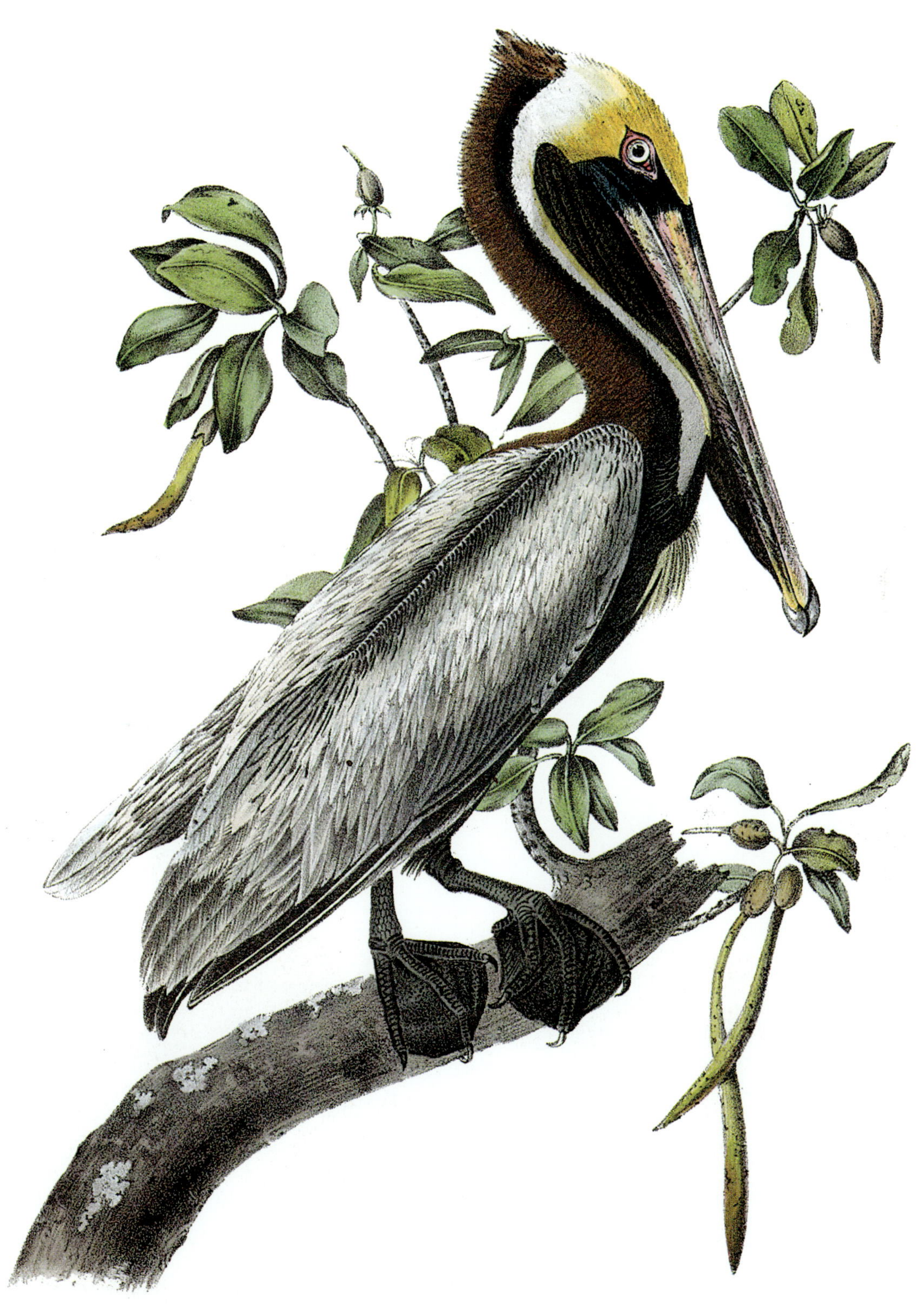

125. Brown Pelican

Adult Male

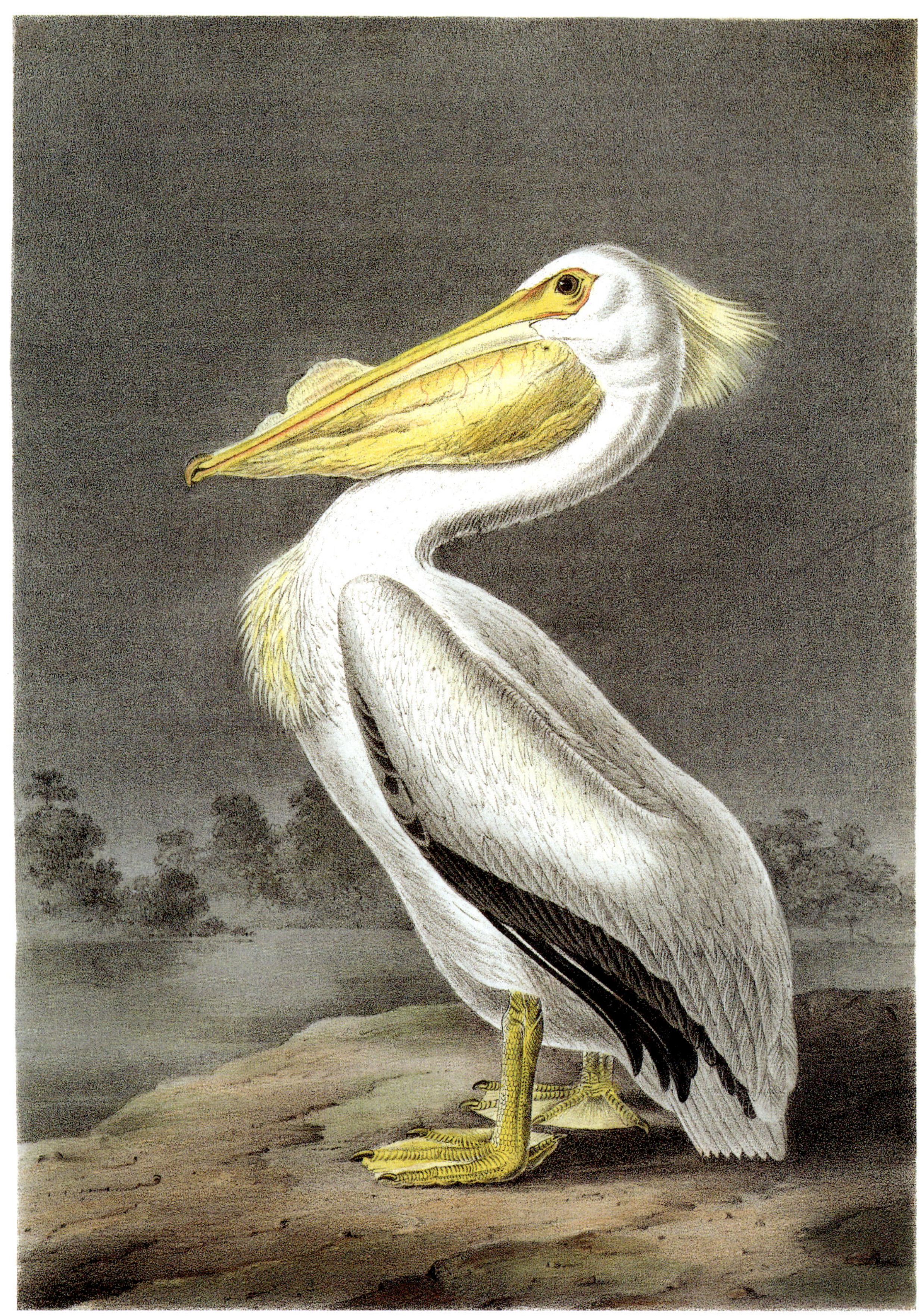

126. American White Pelican

Male

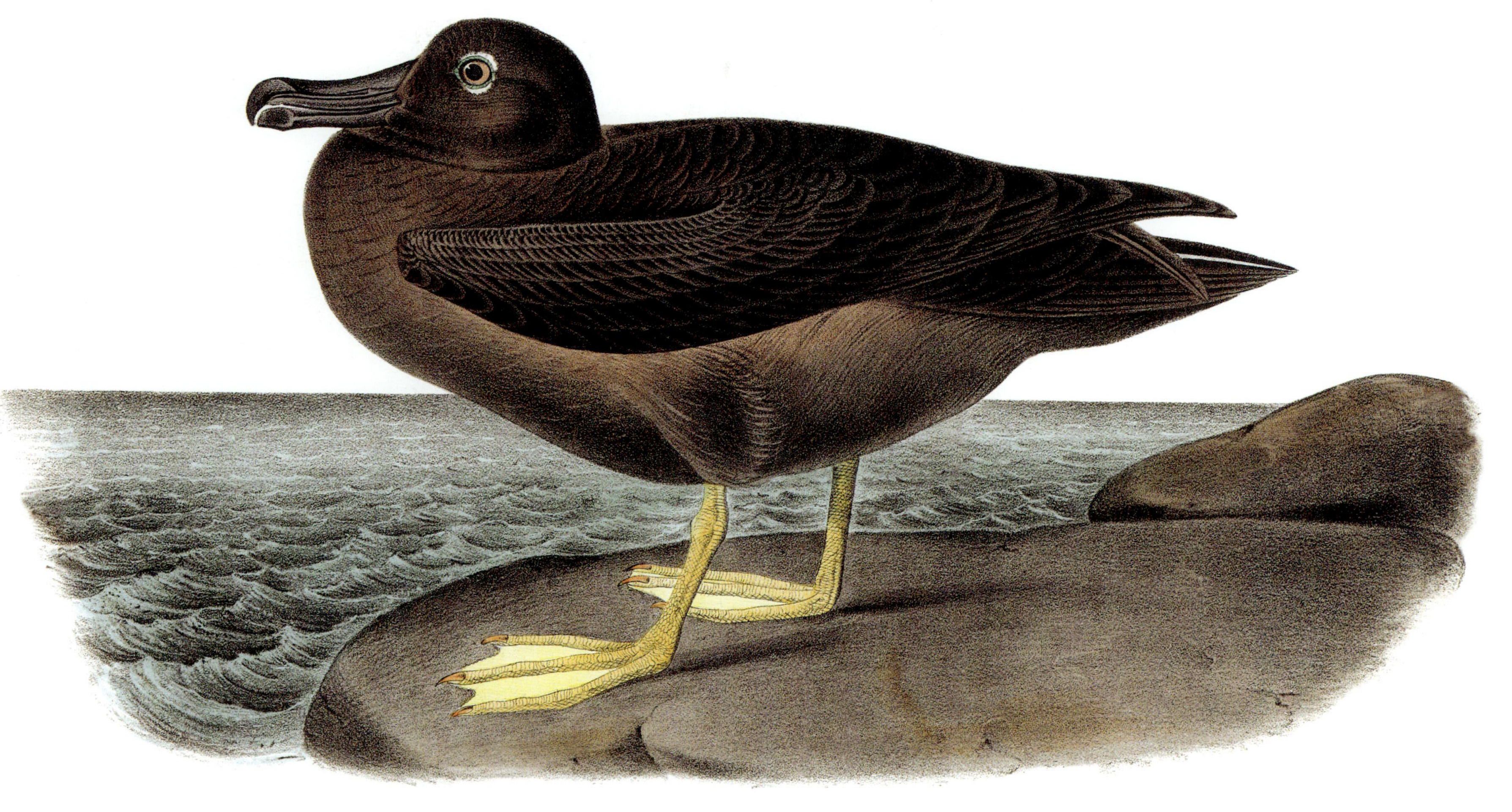

127. Light-Mantled Sooty Albatross

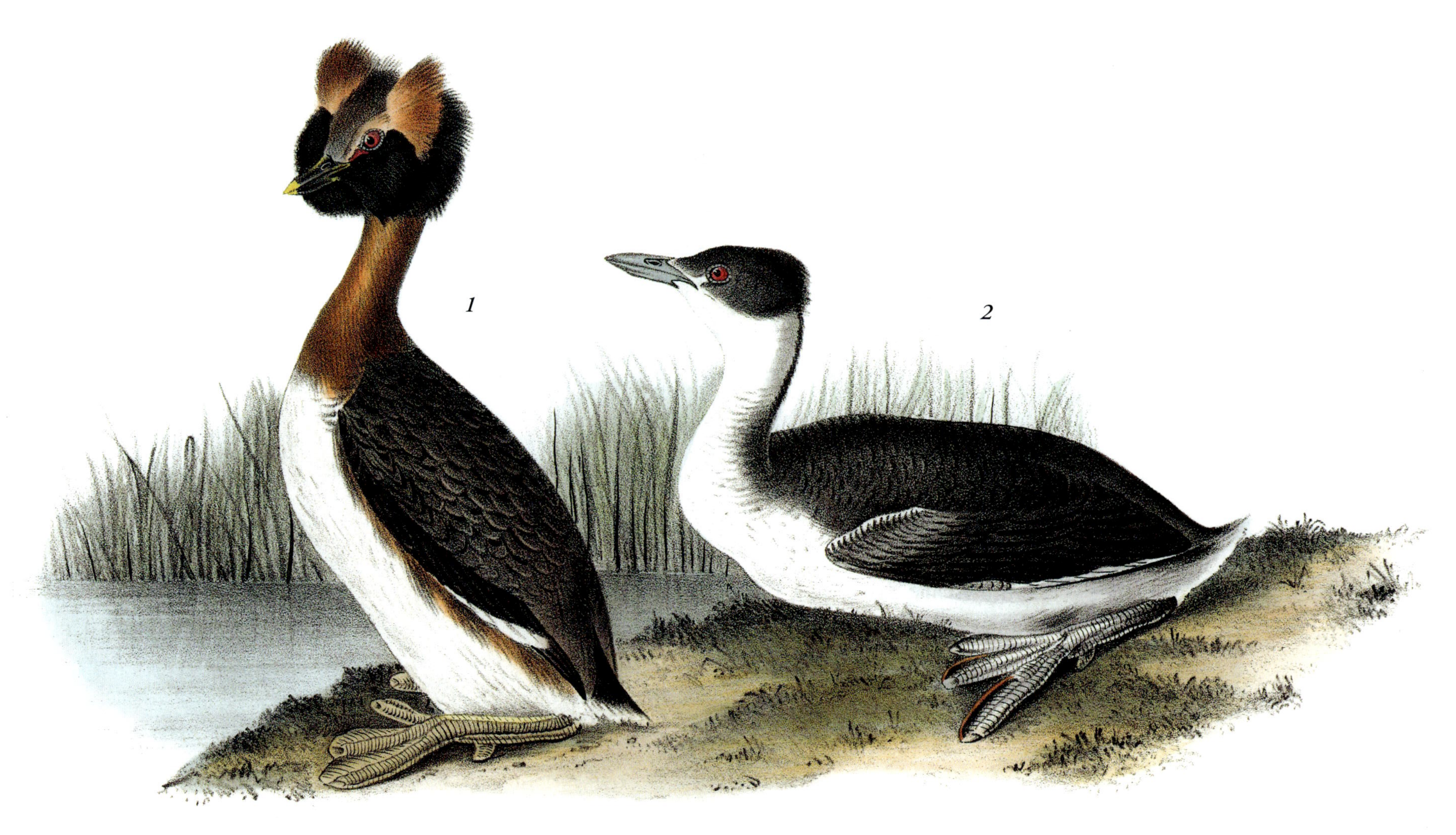

128. Horned Grebe

1. Adult Male 2. Female in Winter

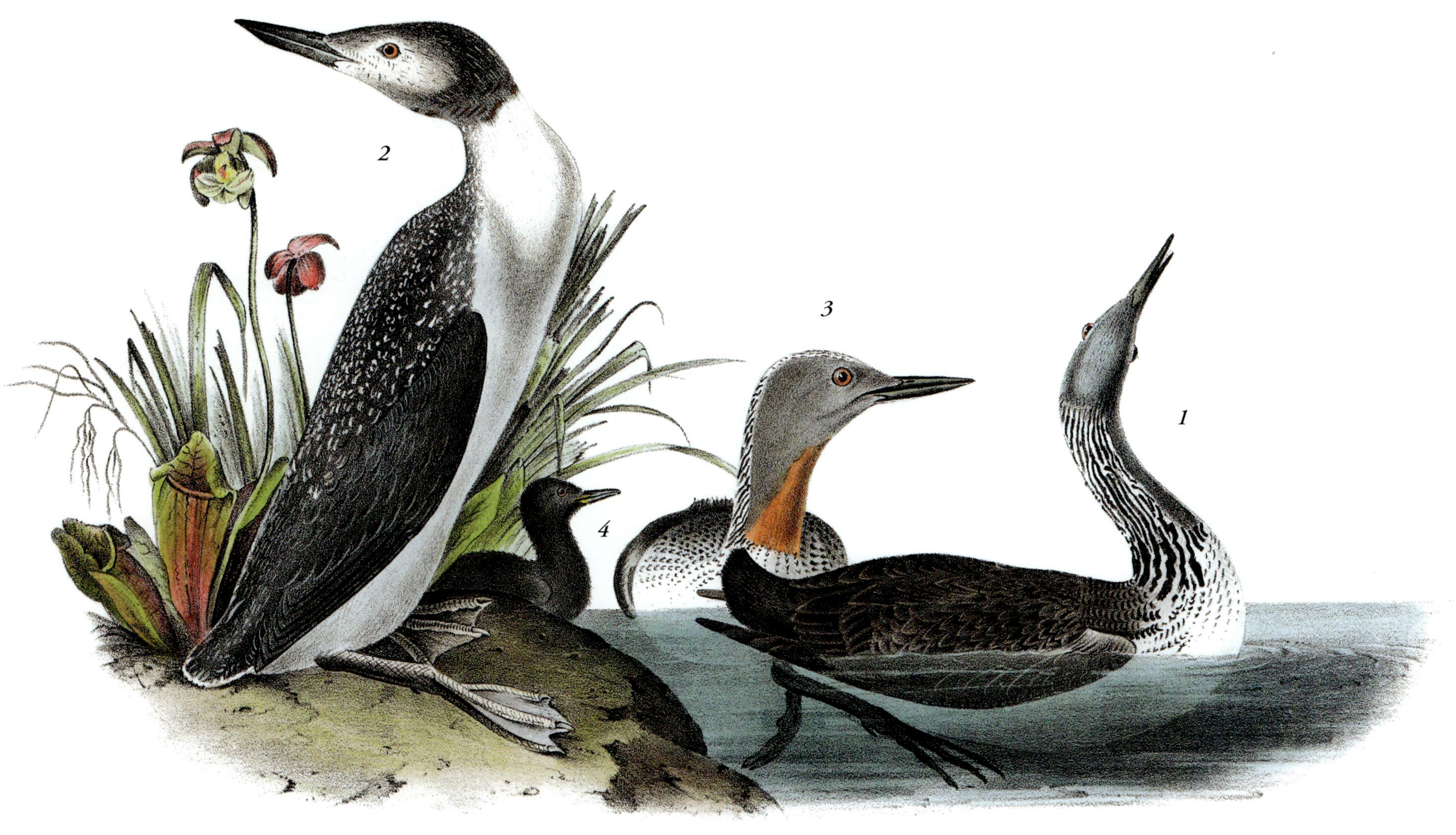

129. Red-Throated Loon

1. Male, Summer Plumage 2. Male, Winter Plumage 3. Female 4. Young

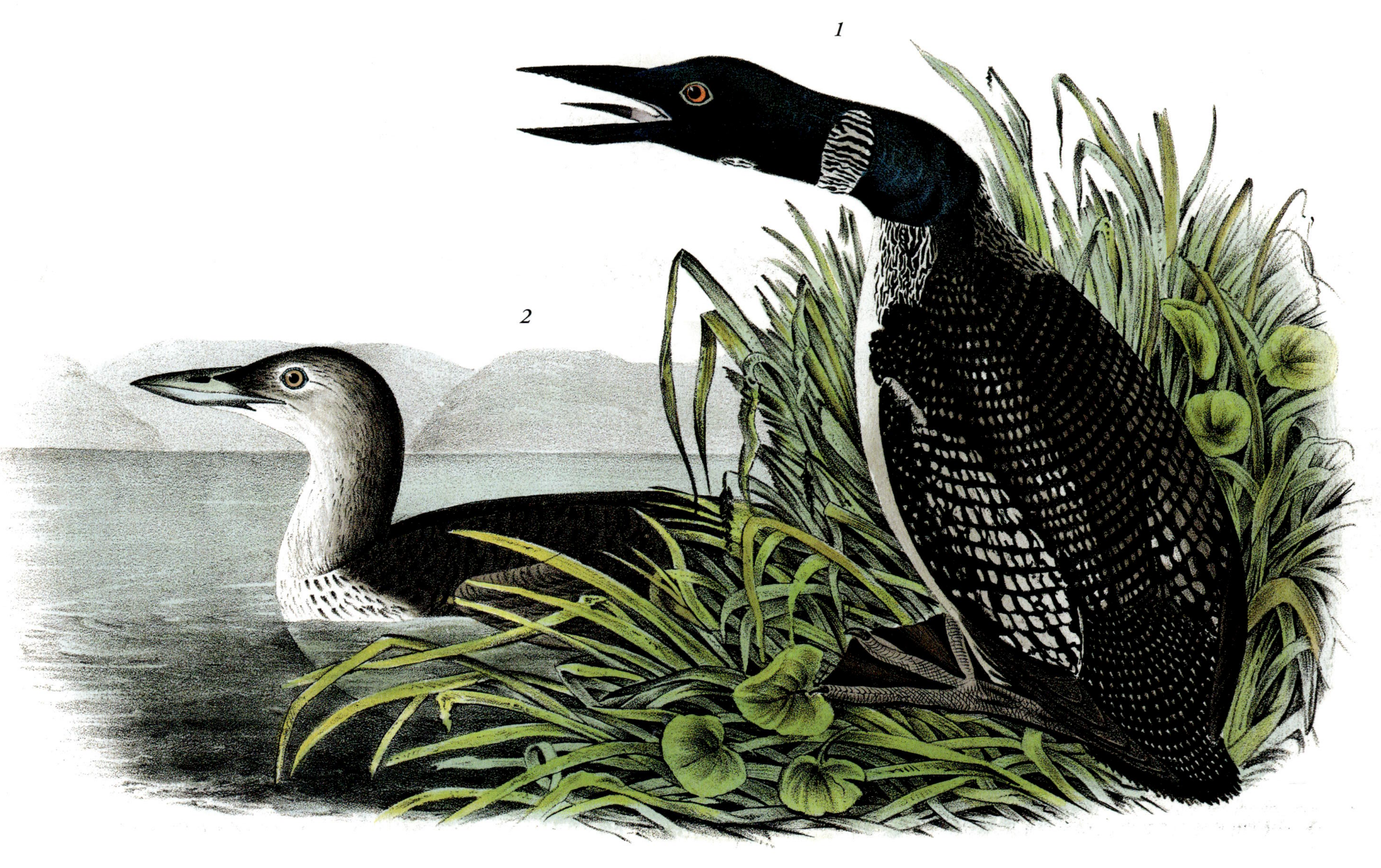

130. Common Loon

1. Adult   2. Young in Winter